Praise for *The Virility Paradox*

"*The Virility Paradox* by Charles Ryan, MD, is a marvelous, timely, and fun book about testosterone and behavior in men and women. This is a *must read* for parents, teachers, doctors, psychologists, and everyone interested in gender."

**—Louann Brizendine MD, neuropsychiatrist,
author of *The Female Brain* and *The Male Brain***

"Charles Ryan has written a unique, easy-to-read medical commentary on the paradoxes of testosterone. Anyone wanting to learn about the scientific basis of puberty, crime, behavior, or prostate cancer will want to read this fascinating book written by one of our nation's most outstanding academic physicians."

**—Stanley Prusiner, MD, director of the Institute for
Neurodegenerative Diseases at University of California,
San Francisco, Nobel Laureate in Medicine, and author of
Madness and Memory**

"Fascinating! Ryan is a doctor and professor but also a natural scientist . . . *The Virility Paradox* is a book of remarkable scope, for this hormone plays a role in every human drama—in sickness and health and in aging; in sex and power; morality and politics; in how we think and how we behave. With unforgettable stories to complement his extensive scientific research, Ryan has us not only rethinking all our ideas about masculinity but what it is to be human."

**—Jena Pincott, author of *Do Chocolate Lovers Have Sweeter Babies:
The Surprising Science of Pregnancy* and *Do Gentlemen Really
Prefer Blondes?: The Science of Love, Sex, and Attraction***

"A terrific read. Dr. Ryan writes with grace, wit, and insight on a topic that touches on nearly every aspect of our society. Highly recommended."

**—Dr. Kelly Parsons, author of *Doing Harm* and *Under the Knife*
and professor of urology at the University of California, San Diego**

THE VIRILITY PARADOX

THE VIRILITY PARADOX

The Vast Influence
of Testosterone on Our Bodies,
Our Minds, and the
World We Live In

Charles J. Ryan

BenBella Books, Inc.
Dallas, TX

BenBella Books, Inc.
10440 N. Central Expressway, Suite 800
Dallas, TX 75231
www.benbellabooks.com
Send feedback to feedback@benbellabooks.com

Printed in the United States of America
10 9 8 7 6 5 4 3 2 1

Library of Congress Cataloging-in-Publication Data Control Number: 2017053846
ISBN 9781944648565 (trade cloth)
ISBN 9781944648572 (electronic)

Editing by Alexa Stevenson and Lisa K. Marietta
Copyediting by Brian J. Buchanan
Proofreading by Lisa Story and Michael Fedison
Indexing by WordCo Indexing Services, Inc.
Text design by Aaron Edmiston
Text composition by PerfecType, Nashville, TN
Cover design by Pete Garceau
Jacket design by Sarah Avinger
Printed by Lake Book Manufacturing

Distributed to the trade by Two Rivers Distribution, an Ingram brand
www.tworiversdistribution.com

Special discounts for bulk sales (minimum of 25 copies) are available. Please contact Aida Herrera at aida@benbellabooks.com.

To my dad, Tom Ryan, whom we lost suddenly in 2017. Thank you for sharing your love, wit, wisdom, and all of the best of you.

A special warm acknowledgment to Hudson Perigo for her patient mentorship, technical savvy, and persistent commitment to excellence that made completion of this book possible

CONTENTS

INTRODUCTION

L et us begin at the end—the end of a life.

Although his death certificate doesn't say so, my patient Jimmy died because of the irrepressibility of testosterone. I watched it happen slowly, over years, and did my best to stop it.

Jimmy was in his early seventies and, like me, was a big, tall guy with an Irish surname who wore his heritage proudly. During the four years he was under my care, we'd bonded over that and other commonalities ranging from raising daughters to the San Francisco Giants. We became close, as is common in my line of work if you choose to let it happen (and not all of us do). After his death his wife, Barbara, gave me a couple of his cashmere sweaters. They were beautiful, and for a while I made a point of wearing them to work on days when I might run into her while she worked at the hospital gift shop. But truthfully, not only were they a bit too big, but I also felt constricted by the grief they carried, and I eventually donated them away. I guess there are limits to that closeness.

My last moments with Jimmy occurred less than an hour before he died. Barbara had called while I was flying home from a work meeting, and I listened to her voice mail while driving back from the airport. Her soft voice was full of pain; she was all alone with Jimmy and knew this was the end. Could we talk? she asked.

I called her back and arranged to come by. Their apartment was on my way home, and I appreciate that the occasional house call is good for the soul, mine included.

I parked on the steep, sloped street and made my way up the staircase to the door of a classic San Francisco Victorian, where Barbara greeted me with a tight hug. She's one of those short, energetic women for whom the word "feisty" seems to have been invented, but her worry, fear, and exhaustion were palpable.

Most cancer deaths now occur at home, and while home care eliminates a costly and needlessly sterile hospital death, it also lessens the chances that a doctor will be at the bedside at the end to support and comfort patients and their families. This insulation between doctors and dying, though it may protect us doctors from burnout, also removes us from the raw corporeality of witnessing death—an experience that can inspire us to fight even harder for the living.

Jimmy was in the bedroom, in an adjustable hospital bed that had been placed next to the bed he'd shared with Barbara for decades, close enough to allow the couple to touch. One look confirmed that although his heart was pumping and his lungs were taking in air, Jimmy had crossed into a place from which he would not return. He was breathing rapidly and barely responsive. His previously muscular body had been reduced to bony protuberances and wasting muscles, all of it tinged a sickly yellow, and the ammonia-like smell of his breath spoke of the necrosis raging inside him. He didn't appear to be in pain, but he didn't appear to be at peace, either.

I stayed for about an hour, examining Jimmy, speaking to him in a reassuring tone, and holding his hand while Barbara and I discussed the next steps. Medically, there wasn't much to do—a dose of morphine to ease the labored breathing, little more. It wouldn't be long. She had questions—How much time did they have? Was he in pain? What about medication?—and I did what I could to answer her. But mostly we just sat.

Soon, their daughter arrived and Jimmy seemed more comfortable. Sensing that it was time for me to go, I quietly said goodbye, leaving the intimacy of Jimmy's remaining moments to his immediate family. If they needed anything, the hospice team was just a few minutes away.

I left at about 7:15; Jimmy died just after eight o'clock.

I still see Barbara from time to time, and I often think about the fact that, if not for prostate cancer, she and Jimmy would likely have enjoyed many more years together. Their story is not an anomaly: cancer kills about 600,000 Americans every year, and around 30,000 of those deaths are from prostate cancer. This is the cancer I treat—and as I care for men like Jimmy and study their disease, I am struck by the biological, evolutionary, and even philosophical uniqueness of the substance that, one way or another, contributes to so many of these deaths. I am referring, of course, to testosterone.

This book is about the breadth of testosterone's effects on the human body—we'll call those effects *virility*. Testosterone is often thought of as the "male hormone," but while it is indeed responsible for the secondary sex characteristics we associate with men (muscle growth, deepening voice, body hair, and so on) as well as stereotypically "manly" attributes like physical aggression and higher levels of sexual arousal, in fact testosterone is at work within all of us, all the time, and has been since well before we were born. On an individual level it influences our decisions and informs our drives to mate and to survive, and testosterone has shaped our evolution and much of our progress as a species. Yet, fueling prostate cancer is not the only black mark on testosterone's record: it is also responsible for a host of our more sinister behaviors and some of humanity's darkest moments. But can we have the good without the bad?

I am fortunate enough to work as an oncologist and a professor at one of the country's major centers for cancer research and treatment, the University of California–San Francisco. I've focused predominantly on treating patients with prostate cancer in its advanced stages, along with researching the condition and developing new treatments. Prostate cancer is a complicated disease and one that can present and progress very differently from patient to patient. One in six American men will get it, and though some will never suffer from symptoms or need any treatment at all, others will die rapidly and dramatically despite our best efforts, while many more will fall somewhere in the middle. Treatment, and the response to it, can be just as variable: primary among the challenges is the fact that while prostate cancer is not *caused by* testosterone, the disease is *driven by* it, and treatment often involves medically suppressing this hormone. Some men are cured with relative ease, while others are overtreated and experience negative side effects; others, in an attempt to avoid these side effects, are undertreated. Finally there are the cases like Jimmy's, in which the effects of testosterone resurge to the point of lethality as the cancer evolves and worsens despite treatment. Given all of this, it is perhaps not surprising that there is a fair amount of confusion about prostate cancer among the general public.

And there's another problem. Because so many other biological processes rely on testosterone, suppressing the hormone as part of cancer treatment can introduce a whole new set of issues. Navigating this dilemma with my patients and working to develop therapies that make this navigation easier is my life's work, and the source of my fascination with testosterone and its effects.

Before we go on, let's review a topic you likely covered in high school biology: the endocrine system. Hormones (testosterone, estrogen, progesterone, cortisol, thyroid hormones, etc.) are made in one or more of the body's glands—organs designed to secrete vital chemicals into the blood. In the case of testosterone, these glands are the testicles in men and the ovaries in women (though for both sexes a small amount is also produced in the adrenal glands). Hormones are released from glands into the bloodstream and from there can enter tissues and exert their effects by binding to a *receptor*. Imagine the receptor as a sort of molecular catcher's mitt. Only cells with the proper receptors can "catch" a specific hormone, and in the case of testosterone, this is the *androgen receptor* (AR). When enough ARs are present in a given tissue, testosterone binds to the receptors and the action begins. The effect testosterone has can vary based on the number and reactivity of these receptors, which can vary due to genetics (among other things); and if a tissue has no ARs or not enough of them, testosterone will have no effect at all. This is why testosterone affects some parts of the body (your hair, your moods, or your muscles), but not others (your eyes or lungs, for example).

Another important player in the endocrine system is the brain. Not only does the brain have receptors itself for various hormones (androgen receptors among them), but a small structure in the brain called the hypothalamus also operates as a sort of thermostat, regulating hormone levels to keep them in balance. When a particular level gets high, the hypothalamus responds by sending a signal to the corresponding gland instructing it to stop secreting the hormone in question. Likewise, when the level gets low, the hypothalamus instructs the gland to step up production.

With testosterone it works a bit like this: testosterone is made in the testicles or ovaries in response to stimulation from the brain, and from there it flows through the blood and finds tissues with androgen receptors, where it can bind and turn on its effects. In muscle tissue, for instance, the effect is the muscle getting larger and stronger. In the skin, testosterone may lead to hair growth or hair loss, depending on the location and the individual (if the cell that receives testosterone is in the skin of the scalp, it may suppress hair growth, but if that cell is in the skin of the face, it may make a beard grow—more on this later). You get the idea.

Testosterone is a member of the *steroid* family of molecules, which also includes estrogen, the primary female sex hormone. (When you hear about an athlete taking steroids, it is likely some form of testosterone or a related molecule, even though the term *steroid* covers a range of hormones.) One detail I've always found fascinating is how similar all the hormones are to one another, despite their dramatically different effects on the body. Testosterone, for example, is made up of nineteen carbon, twenty-eight hydrogen, and two oxygen atoms (its molecular formula is $C_{19}H_{28}O_2$), whereas estradiol, or estrogen, has eighteen carbon, twenty-four hydrogen, and two oxygen atoms ($C_{18}H_{24}O_2$). The difference between estrogen and testosterone is just one tiny carbon and four hydrogen atoms, yet the effect of those differences governs how you look, how you think, and more. This isn't some sort of molecular coincidence, either; estrogen is *made from* testosterone through one quick chemical conversion. (Seems like there could be an Adam and Eve metaphor here, but I'll refrain.)

Let's exchange our ball-and-mitt metaphor for one a bit more accurate and nuanced—a key and a lock. Imagine that the minute differences between estrogen and testosterone are like notches on a house key. The receptor, then, is the lock—androgen receptors are unlocked by testosterone, estrogen receptors by estrogen. In most cases the key turns smoothly and the door swings open, but sometimes locks get stuck and sometimes keys get jammed. Some doors have many locks, some doors don't have any, and some locks get used a lot, others only rarely. One thing is certain: the wrong key will never open the wrong lock, no matter how hard you try, and the same is true of hormones and receptors. Testosterone needs to bind to an androgen receptor in order to take effect. This concept is the foundation from which we can start to explore the intricacies of the testosterone system, and these are two of the forces at work in what I call the *virility triad*. The virility triad is the system by which testosterone expression is governed through relationships between testosterone levels in the blood, variations in androgen receptors, and the influence of fetal testosterone—each of which will be discussed more later in this book.

Testosterone continues to fascinate me in part because, like our species as a whole, it is driven to *win*. This molecular system seems to have its own survival instinct; we might suppress it temporarily, but in the long run it is irrepressible. The hormonal therapy Jimmy received eventually stopped working because prostate cancer did what cancers do—harnessed the molecular survival instincts within its host. These molecular instincts emerge in other arenas of life as well. If we could find a way to overcome that irrepressibility, we might cure prostate cancer, but who knows what damage we would do by eradicating testosterone entirely?

This is the paradox of virility. The story of testosterone and its expression in humans has two sides, one beautiful, one ugly. To our benefit, virility shapes our desire to explore, build, survive, and procreate—in both men and women. It gives us strength, brings us together for mating, and so drives our evolution. If not for testosterone, our species would have died out long ago, and virility and its effects permeate the art, literature, and fellowship of every known culture. Without it, the world would be a drastically different place.

Yet, there's another side. Scientists have found associations between virility and violence, crime, poverty, and unstable relationships. Recent experimental data shows that high levels of testosterone can negatively affect a person's capacity for compassion, generosity, and empathy. When the delicate balance of our hormones is disrupted, the result can be any number of diseases and disorders, and testosterone can fuel processes with the power to destroy us.

Many of the examples we'll discuss in this book illustrate what happens when testosterone levels fall outside standard ranges, and through the stories of real people, we'll explore the connections between testosterone and dementia, autism, sexual aggression, menopause, athletic ability, crime, fatherhood, and empathy. Looking at outliers can give us insight into how testosterone works in the bodies of all of us, and suggest how we might use this knowledge to improve our lives.

I am not a psychologist or evolutionary biologist; I am a clinician who cares for living patients and, as such, I approach clinical problems as a confluence of biological, social, and environmental factors. That's how you treat cancer, and I think it may be a useful way to analyze and interpret the effect of this very interesting molecule on the world. My investigation into this powerful hormone will begin with prostate cancer patients like Jimmy but will fan out to include men, women, and children from many walks of life, each of whom has something to teach us about how testosterone informs the human experience. It is my hope that as you learn about the ways in which testosterone drives our bodies and our behaviors, you will also gain a greater appreciation for the ways in which we humans have the power to control, manipulate, and even overcome our biology.

Part I

THE CHEMICAL THAT BINDS US

Chapter One

THE LITIGATOR'S METAMORPHOSIS: FROM COMPETITION TO COMPASSION

At sixty, my patient Aaron was enjoying an enviable life as a father, a new husband, and a successful Los Angeles lawyer. I liked him immediately, although whether this was because or in spite of his chest-popping confidence was hard to say. He was gregarious, his personality part star quarterback, part comedian, his conversation always peppered with self-deprecating lawyer jokes.

The sole hiccup in his otherwise healthy life had come a few years earlier when he had surgery for early-stage prostate cancer. For many patients, this is the beginning and the end of their prostate cancer story; the cancer is gone and life goes on. Because prostate surgery has no effect on testosterone, Aaron's virility was undimmed, his testosterone level measuring right in the normal range for a man of his age.* He continued to stride confidently through life, assuming he had decades of vitality ahead of him. Aaron was used to winning—cancer was simply another vanquished opponent.

Unfortunately, that sense of invincibility came crashing down a few months before I met him, when a routine blood test revealed that his levels of PSA (prostate-specific antigen) were rising rapidly, a clear sign that the cancer persisted somewhere within him. From that moment on, we knew we couldn't cure Aaron's cancer. Luckily, it was still very treatable, with a prognosis measured not in weeks or months but years—hopefully a decade or more. We'd control the cancer's spread with what is called "hormonal therapy." It's a misleading term, as rather than providing additional hormones, as you might expect, the therapy involves suppressing one: testosterone. The medication I prescribed (a shot given every three months) would reduce Aaron's testosterone from about 400 nanograms per deciliter to somewhere between 25 and 50 ng/dL. (In some cases, levels can fall all the way to zero.) Suppressing his testosterone is the first step in controlling the disease.

Testosterone fuels the cancer cells threatening Aaron's life, but it also fuels Aaron himself, affecting everything from his libido to his mood, appearance, and even the decisions he makes on a daily basis. Thus, in a way, patients with very low testosterone levels due to hormonal therapy (there are over a million of them) are a living laboratory for study of the hormone's effects on the bodies and minds of men—similar to what is called a "knockout mouse" model, an experiment in which a specific gene is eliminated in a strain of mice, allowing us to understand the gene's function by observing the changes that arise in its absence. Hormone depletion can hold cancer at bay for a long time, but at what cost to Aaron and others like him? On the other

---

*Testosterone is measured in the bloodstream. The measurement is sometimes referred to as *serum testosterone level.*

hand, could there possibly be benefits to low testosterone that extend beyond cancer treatment?

THE WINNER EFFECT

Testosterone spikes when we win—at just about anything. This is known as the "winner effect," and it was one of the earliest and most important observations linking behavior to testosterone. These spikes in testosterone occur in competitive settings of all types, from sports and business to dating, hunting, and even chess. In both men and women, testosterone increases feelings of confidence and assertiveness. It also spurs the release of dopamine, a powerful feel-good chemical.[1] What's more, not only does this winning burst of testosterone make us feel dominant and primed for further competition, it also ensures that this competition will be even *more* rewarding in the future: the higher levels of testosterone following a win stimulate the production of more androgen receptors in the brain,[2] in essence making more cylinders for this fuel to drive. It is what we call a "feed-forward" system. The spike in testosterone is both an *effect* of the victory—one underlying its thrill—and a *cause* of further testosterone-driven thoughts and actions.

Aaron is an embodiment of a persistent winner effect. It transformed him. During his visits to my office he's slowly opened up to me, reflecting on how he became what he calls a "card-carrying He-Man."

As a child growing up in 1960s Chicago, he stuttered and was overweight. Classmates called him "Porky Pig," and at home he suffered the insults of yet another put-down artist: his father.

"Dad looked at the stuttering as a sort of performance failure," he told me. "I guess I was an embarrassment to him."

Aaron grew up an angry kid; there were fights on the playground, with his brother, and even in the classroom, which led to frequent detentions and more wrath from his father. He yearned to escape, and after he'd worked his way through college as a taxi driver, his golden opportunity arrived, in the form of admission to one of California's top law schools. This is where his transformation began, one that epitomizes the feed-forward nature of the winner effect in particular and the testosterone system in general—a system in which success breeds success—at the molecular, personal, and societal level.

Aaron took on the law, California, and his future with vigor and enthusiasm. With the move West, away from his past and his father, he flourished. The stutter long gone, he was easily the smoothest talker in his class. His athletic ability also blossomed, and he became a regular at the local gym and began competing in club basketball tournaments. Well built, with soft blue eyes, he attracted women drawn to his eloquence, athletic body, and confidence in the law school's mock trials. ("I was a womanizer," he confesses. His voice says he is reminiscing; his eyes betray a boast.)

After graduation, Aaron put his brimming confidence to work in the public defender's office, managing huge caseloads and priding himself on speaking out for those with no one else to speak for them. Judges, the police, and other lawyers took notice, and the more he won and the more fame he gained, the more fearless he felt. Ready for a new challenge, Aaron left public defense and started his own firm with a colleague, continuing to take on cases nobody else wanted: suing police departments, municipalities, and other guardians of public safety for "acts of irresponsibility" that left people injured or dead. Meanwhile, he and his law partner competed for everything: the toughest cases, the biggest payouts, the most publicity. On weekends they'd attack one another on the basketball court, in pickup games or one-on-one.

Aaron became known for his eloquent, often artful closing arguments, which often produced six- and even seven-figure awards for his clients. He commanded respect; moreover, his track record and confidence instilled a fear in opposing counsel that carried over to subsequent cases. In the late nineties, he sued a local police jurisdiction on behalf of the family of a young man who had died in their custody—and won, setting a state record for the largest punitive award of its kind. In the courtroom, he was a master of expressing the anger and frustration of others who had been wronged, all while keeping his own anger under control. Although he admitted to occasional pangs of moral ambivalence, or even sympathy for the other side, he trained himself to suppress these emotions in order to accomplish his goals. He had honed a warrior mentality, and it was paying off. At various points in our conversations he has referred to his work in terms of the NFL, Ancient Rome, and the military, all metaphors of conquest, competition, and victory.

This, then, is the winner effect. Winning in head-to-head competition—in the court of law or on the basketball court—would have consistently boosted Aaron's testosterone levels, making him a more confident and aggressive competitor, making him less risk-averse, and increasing his sensitivity to testosterone in the future in a self-perpetuating loop of victory and payoff. Although lawyers on the whole don't have notably higher testosterone levels than the general population, *trial* lawyers do have significantly higher levels than do non-trial lawyers, regardless of their sex.[3]

But what does the winner effect really do for us? It's a safe bet that nature didn't put it there just for lawyers. Most likely, the spikes in testosterone we call the winner effect evolved by providing a *selective advantage*—an anthropological and evolutionary term for a trait that makes individuals more likely to mate and breed. Beyond the individual, this advantage would eventually have an impact on the species as a whole . . . and on our societies.

Anthropologist Ben Trumble, at Arizona State University, has made the study of population-wide variations in testosterone and other hormones his life's work. His laboratory, so to speak, is the living environment of the Tsimane people (pronounced *chi-mah-nay*) in the Amazon region of Bolivia.[4] This isolated group of about nine thousand indigenous people survives on subsistence farming, foraging, and fishing in the wildest parts of the rain forest. Studying their lifestyle is the closest we're likely to get to observing the world of the hunter-gatherer ancestors from whom we evolved. Trumble's work suggests the Tsimane can serve as a model for how spikes in testosterone during hunting, sports, and other forms of competition (litigation, for instance) correlate with reproductive success. In traditional hunter-gatherer societies, returning from the hunt with a kill for the village not only filled nutritional needs, but it also made the victorious hunter attractive to women—and, because of the spike in testosterone in the wake of "winning," increased his sex drive. More sex and thus more children provides another feed-forward signaling loop, as men with high testosterone go on to produce high-testosterone offspring, imprinting the winner effect on our evolution. This is consistent with our drive for *provisioning*, or the need to provide for our mates, our offspring, and ourselves. The winner effect's influence on provisioning defined how our hunter-gatherer ancestors survived, thrived, and created the next generation.

While many of us simplify evolution into the concept of survival of the fittest, it's really about *reproduction* of the fittest—after all, the point of survival, from an evolutionary standpoint, is to survive long enough to reproduce and pass on our genes. In this case, the winner effect makes successful hunters more fit *and* more attractive as mates. This is one illustration of Darwin's theory of *sexual selection*, which is distinct from the theory of *natural selection*. In sexual selection, the selection is made not by fate or nature but by the females of the species as they choose their mates.

While sexual selection is a major factor in shaping societies, survival of the fittest remains a powerful force in the evolution of the species, although I propose that our definition of "fit" in this context needs a reboot. In every society, better providers are seen as being more attractive; the idea is that those who are making the sexual selection (i.e., females) are making that selection based on a male's perceived fitness. While this basic fact hasn't changed, our definition of what makes a better provider has. As most cultures evolved, virile physical attributes such as high muscle mass, speed, and strength were traditionally considered attractive because they signaled a mate who not only would be a successful provider but who also would pass down genes that would make subsequent generations healthier and better able to protect themselves from threats of various kinds. Although many women still see these attributes as attractive, today we have more abstract concepts of fitness; instead of hunting prowess, higher income is a better indicator of a man's ability to provide for himself and his family, and big muscles are no longer a sign of strength so much as they are evidence of the kind of financial freedom that allows a person to spend—and pay for—time at the gym. For some segments of the population, physical strength is a luxury, not a necessity, and it is certainly not as closely associated with survival as it once was. The savannahs and rain forests of our ancestors have given way to other jungles, mostly concrete ones, and it is there that testosterone and the winner effect meet Wall Street.

WINNING IN MODERN SOCIETY

Behavioral economist and former hedge fund investor Ben Coates and his colleagues at Cambridge University studied associations between testosterone levels and success in the day trading of stocks. Like all good experimental methods, his was fairly simple: look at testosterone levels at various times during the day and compare them to the success of the investments made. Day trading, as the term suggests, involves the rapid movement of stocks in and out of the market, usually within hours or even minutes. This fast-paced, high-stakes environment mimics a sporting competition or, perhaps even more precisely, gambling at a casino. Since every rise in testosterone "primes" the positive feedback loop, encouraging more risk-taking, a trader who has a lucrative morning would, in theory, have a more lucrative afternoon than a trader who did not experience the early surge of testosterone that encourages him to take more risks as the day progresses.

Coates set out to test this hypothesis in London's version of Wall Street, known as the City. Coates and his team followed seventeen day-traders at a firm for eight consecutive days. His subjects (all men) were trading real money and in large amounts; the value of each trade varied from £100,000 to £500,000,000 (roughly equivalent to between $120,000 and $650,000,000). Tracking each trader over time allowed each subject to be his own control—that is, the researchers could compare a trader's average day to the days on which he fared better or worse. The team's prediction was that, consistent with the winner effect, testosterone levels would rise on the more successful days and decline on the less successful days.

Coates's findings were as follows: First, on days when the traders' testosterone (tested at 11 AM, before most of the trading was to take place) was above their average daily value, they made more money. Second, on the lower-testosterone days, they lost money. On days with above-average testosterone readings at 11 AM, the profit range was between £700 and £800, compared to zero profit on the days the team recorded below-average testosterone levels.[5] (The experiment that should have followed but didn't would have been to send traders home at 11:01 AM if they had below-average testosterone levels!)

Now, nobody is going to suggest that success in the stock market depends on testosterone levels, or that those with lower testosterone can't be profitable traders. The value of this study lies in what it tells us about testosterone, and what it shows us about how the winner effect can manifest in modern society. Whereas Darwin's description of a "fit" human might not conjure images of a successful day-trader, the concept is the same, and it's not a stretch to make connections between how testosterone influenced both sexual and natural selection thousands of generations ago and how it functions in communities today.

IS THERE A LOSER EFFECT?

If the winner effect works as a positive feedback loop, ultimately increasing an individual's testosterone levels over time, then does chronic losing lead to persistently lower testosterone? If so, could this occur on a mass scale and affect entire populations, such as those trapped in or fleeing war zones, living in poverty, or enduring institutionalized oppression?

My own father's demographic may illustrate the loser effect on a generational scale. He was fond of reminding me that he was born in 1935, smack in the middle of the Great Depression, a year with the lowest relative birthrate on record in the United States. He claims this circumstance led to all manner of economic advantages, such as easier admission into college and medical school (he is a retired oncologist and internist) because of a shrunken pool of applicants. This situation resonates with what Warren Buffet, born a few years earlier in 1930, has cited as the secret to his success: being born in the right place (the United States) and at just the right time.*

I've often wondered about the root of this phenomenon. How exactly did economic downturn during the Great Depression affect the nation's birthrate? Was it because men and women were separated as the men left home to look for work? Were couples more diligent

---

*My dad was fundamentally correct, but his facts need to be updated. It turns out that 1935 had the lowest birthrate on record *at the time*, with 18.5 live births per 1,000. After World War II the rate shot up during the baby boom, then stabilized at about 25 live births per 1,000. The big change came in the early 1960s, when it dropped to its current rate of between 14 and 16 live births per 1,000, largely the result of the availability of effective contraception.

about using whatever methods of birth control they had access to? Or was there some biological process at play—perhaps something like the reverse of the winner effect? Could the wave of despair that prevailed in the 1930s have led to lower testosterone levels across the nation, resulting in higher rates of impotence and lower levels of libido, in turn leading to less frequent copulation and, in the end, fewer babies?

While it is not feasible to measure the average testosterone of an entire population, we can learn more about testosterone suppression in other ways. In 1994 a group of researchers led by James Dabbs of Georgia State University took to the pitch to study pre- and post-game testosterone levels in the saliva of male fans watching the TV coverage of the final of that year's World Cup tournament, between Brazil and Italy.

Brazil won the match, and within minutes after the game, eleven of the twelve Team Brazil test subjects showed elevated testosterone levels. Most levels increased from their baselines by about 15 to 35 percent, but two of the men experienced an almost 100 percent spike, or twice their baseline level. By contrast, fans of the Italian team experienced a decrease in testosterone levels. *Every one of them.* And the reductions were similar in magnitude to the increases seen in their counterparts. In fact, two Italian fans experienced more than a 50 percent drop![6]

Keep in mind that these numbers were recorded within just minutes of a single loss. Consider some of the chronic stresses many of us face, from economic and emotional worries to life-threatening health issues. Imagine the effect of defeat, subjugation, and deprivation over the long term, and in contexts more serious than a soccer game. These stressors can actually change the body and the brain. Indeed, studies have shown that feelings of defeat suppress the release of testosterone, as do extreme psychological stresses such as army boot-camp training, or the start of a prison term. Could this phenomenon be at the root of the low Depression-era birth rate? We can only speculate.

ENTER TESTOSTERONE, EXIT EMPATHY?

To review, losing is bad and winning is good . . . but does it come at a cost? Aaron has been reflecting on this as he undergoes hormonal therapy for his prostate cancer, contemplating his past and his commitment to the "win at all costs" mind-set.

At one appointment, our conversation turns to the notion that when a person is completely focused on winning, it's easy to ignore the humanity of others. I nod in agreement, reflecting on my own experience that, for all the humanity at its core, a career in academic medicine requires more than a bit of a competitive streak. The presence of a winner always implies the existence of a loser, and since the desire to win requires some degree of indifference to that loser, the trap many habitual winners fall into is seeing other people not as individuals but merely as obstacles on the path to success.

"The courtroom was my gridiron; I was there to win," Aaron proclaims with a pride that makes me sit up straighter and want to compete myself. I notice he's made a fist with one hand and is using it to lightly pound the other, open on his lap. Yet he admits that his inclination to shut out natural feelings of empathy became a problem beyond the walls of the courthouse. Like many prosperous men, the winner effect was a tonic for Aaron in his thirties, forties, and fifties; he was athletic, attractive to women, dedicated to his work, and generally considered successful by all who knew him. But he faced the dark side as well, and overwhelmingly the price he paid came in the area of his ability to relate to others. Aaron's first marriage ended in divorce, and even when he was single, his "womanizing" was not without casualties.

Although no one has sorted out the how and why, data shows that divorced men have higher testosterone levels than do men who are still married or who never married at all, and psychologists have also recently observed that high-testosterone men display a lack of emotional investment in the lives of their children. (For the record, Aaron has since happily remarried, and I do not know a prouder father.) "Life history theory," a major concept in anthropology,[7] poses that there's a biological tradeoff between mating and parenting behavior, and research in this vein verifies that men in "mating mode," as determined by testosterone levels and even the size of their testicles, are less likely to display nurturing parenting behaviors, such as empathy for children and crying babies.

"I'm not proud of it now," Aaron says, "but as I look back on some of my early days, the relationships that mattered most were those I could use to get ahead. I was climbing the ladder, the classic 'kiss up and kick down' kind of guy. You only mattered to me if I knew you could get me somewhere or something."

I tell him I recall having similar conversations in my younger years. This phenomenon is common in medical students aiming to get to the top of the class in order to get noticed for the competitive residency slots in coveted specialties. It's probably an impulse as old as the profession, yet those who end up truly on top, I suspect, are not those who harness and exploit this urge but those who are able to control it.

The relapse of Aaron's cancer has been a wake-up call. He's about to lose his testosterone, and that, too, is going to change his perspective—in ways that may be hard to predict.

One provocative finding from experiments done on moral dilemmas and their relationship to testosterone reveals that "avoiders"— those unable to make the kinds of decisions that involve harming someone to save others—had the lowest average testosterone levels. Put another way, higher testosterone reduces doubt and/or increases confidence in situations where judgment might otherwise be clouded by empathy. This attribute might be useful on the battlefield, on the hunt, or even in the courtroom. Looking at another field entirely, one might also argue that "violating" the human body during surgery requires a surgeon to create and maintain a certain emotional distance from the patient.*

By blocking the emotional impulses that might compromise our reaching a specific goal (shooting the deer or the enemy, slicing through the viscera with a scalpel), testosterone allows us to get the job done. In order to kill during the hunt or in self-defense—or to commit various petty offenses to get ahead in day-to-day life, for that matter—it's advantageous to shut off feelings of empathy, and studies show that this is just what testosterone enables us to do. You may note that I said "enables," not "causes." The word choice is deliberate, to avoid giving this one molecule sole blame or credit for all of our day-to-day interactions. After all, biology is complex. And yet, this effect is a real one, and one that can become problematic when it spills out of the realm of life-or-death situations and into the everyday.

So far we've talked primarily about effects tied to current levels of testosterone in the bloodstream, but it is important to note that much of the influence testosterone has on us takes place well before we are born—in the fifteenth week of gestation, in fact. During this time, testosterone levels spike in both males and females (although higher in males, on average), and this surge coincides with a time of intense brain growth in the fetus. Scientists have discovered a surprising way to estimate the effect testosterone had on an individual in utero: namely, by measuring the ratio of the length of the right index finger (the second digit, or 2D) to the right ring finger (4D).[10] The longer the index finger compared to the ring finger, the higher the numeric ratio (e.g.,

---

*If this were true, we might expect more men than women to become surgeons, and indeed surgery remains a male-dominated specialty by a ratio of three to one. Is testosterone at work here, too?[8, 9] Hard to say, given the many complicating social and cultural factors, but there is no doubt that it remains one of the most competitive specialties in medicine, and those who end up as surgeons are those able to flourish amidst competition, at least in their early careers.

an index finger that is 90 percent of the length of the ring finger has a ratio of 0.9, one that is 80 percent due to a longer ring finger is lower at 0.8—women tend to have index fingers longer than their ring fingers, while in men the ring finger is usually longer). This 2D:4D ratio is an approximate indicator of the amount of testosterone a person's brain was exposed to in that fifteenth week, known as *fetal testosterone*.* The higher the ratio, the *lower* the amount of fetal testosterone exposure—in general, women have index fingers that are the same length or longer than their ring fingers and so higher ratios, while with men the reverse is true. But differences in this ratio within and across sexes have proven to have surprising associations, and the simplicity of this observation has produced a flourish of studies that offer fetal testosterone exposure as an explanation for any number of traits.

In 2013, a group of psychologists in the Dutch city of Utrecht studied how testosterone and the 2D:4D ratio might interact to predict moral judgment in adults.[11] In a version of the classic ethics experiment known as the "trolley problem," subjects were faced with the hypothetical choice of killing one person (a stranger) in order to save a number of others. The test subjects were first administered a shot of testosterone and then presented with the following scenario:

> A runaway trolley is heading down the tracks toward five workmen who will be killed if the trolley proceeds on its present course. You are on a footbridge over the tracks, in between the approaching trolley and the five workmen. Next to you on this footbridge is a stranger who happens to be very large. The only way to save the lives of the five workmen is to push this

---

*Although the 2D:4D ratio is useful, some research has posited that a measurement of the anogenital distance—that is, between the anus and the base of the penis or vagina—is a better indicator of fetal testosterone exposure. As someone who often catches himself trying to guess at the fetal testosterone levels of people I know or see on TV, I'm glad I have the option to look at finger length.

stranger off the bridge and onto the tracks below, where his large body will stop the trolley. The stranger will die if you do this, but the five workmen will be saved.

Test subjects with a high 2D:4D ratio (and who thus had experienced less influence of testosterone on their brain formation) were more likely to make the utilitarian judgment (the ends justify the means; kill the stranger to save the workmen) than the more emotion-driven decision (refrain from intentionally killing someone). Put simply, a shot of testosterone made test subjects with high 2D:4D ratios (low fetal testosterone) more likely to make the kill. In a cruel world, or under fraught conditions, utilitarian decisions are sometimes needed for survival, and in many cases our hormones drive us to make such choices; in the trolley experiment, testosterone allowed the test subjects to suppress the emotional trauma associated with the decision to kill an innocent bystander. By contrast, testosterone administration had little effect on subjects with low 2D:4D ratios (indicating a *high* fetal testosterone exposure). Those whose brains had been exposed to more testosterone before they were born weren't as likely to be influenced by the testosterone boost. The findings suggest that higher fetal testosterone leads to a brain that is "hardwired" to be less susceptible to sudden surges in testosterone, whereas lower fetal testosterone may leave the brain more susceptible to the effects of testosterone fluctuations later in life. This may partly explain why some of my patients suffer more than others when testosterone is taken away. It may also explain why testosterone supplementation is a life-changing miracle cure for some and elicits merely a "meh" from others, as all brains are not equally sensitive to these hormonal fluctuations, and they may have been that way since before birth.

One important detail I haven't mentioned about this experiment is that all of the subjects were women. We know that women naturally have lower testosterone levels on average than men, but how does this affect the way men and women make moral decisions?

When it comes to outlining the differences between the sexes, the common stereotype is that men approach decisions from a utilitarian perspective while women are more likely to act based on their emotions. In the above experiments, when faced with hypothetical dilemmas that stimulate certain areas of the brain, women who have been given testosterone shots were more likely to shift their moral judgments from empathetic toward utilitarian, the latter being more typical of men who are faced with these same questions. Does this confirm the popular notion that the higher testosterone levels in men's brains make them more utilitarian decision makers? I think we can challenge that stereotype by pointing out that it is not precisely sex that determines such things but instead distinct areas of the brain that become chemically activated under different circumstances. In fact, the most current thinking suggests that there may not be a "male brain" or a "female brain" but rather that we all move around on a single brain spectrum.[12]

WILL REDUCING TESTOSTERONE INCREASE EMPATHY?

In my own clinical research program, we've started measuring "empathy quotients" in our prostate cancer patients with the help of a survey developed for use in the study of autism.[13] Our goal is to track signs of empathy as our patients' testosterone levels are reduced as part of their hormonal therapy. Aaron takes this survey every three months.

We hypothesize that as testosterone levels are reduced, the capacity for empathy will increase. Although there isn't yet much published research on what we might call "inducing niceness" by taking away testosterone, it isn't exactly a stretch given what we have just seen about the effects of testosterone in general, and research showing that excesses of testosterone can have the opposite effect.

Aaron started hormonal treatments about nine months ago. His total testosterone is down from his previous 399 nanograms per deciliter to 34 ng/dL, and his PSA level is almost at zero; we have the cancer under control.

While all is quiet oncologically, a psychological transformation is taking shape.

At a recent visit, he asked if "getting emotional" was a side effect of the medication.

"What do you mean, specifically?" I asked.

"Well, I was back in Chicago this weekend, visiting my mother—she's ninety-one. And when I was leaving for the airport, I got very emotional and teared up."

"Is that a bad thing?"

He thinks. His eyebrows crinkle a bit and his lips make a slight pucker. "I guess not."

I tell him I get a lump in my own throat in similar circumstances. My mother also lives two thousand miles away, and every time I say goodbye to her I can't help but think about the fact that one day it will be for the last time.

He notes that he never reacted this way before starting hormonal therapy, and so I ask if he's found himself getting tearful in the courtroom or at work, where it might be inappropriate.

"Not yet," he replies.

I sense uncertainty in his response, as if he's no longer sure what the future holds for him.

These emotions may feel strange to Aaron, but I would argue that a lump in your throat as you kiss your ninety-one-year-old mom goodbye is a healthy expression of normal human emotion. In Aaron's case, taking down the wall of testosterone may not be a bad thing.

I believe Aaron welcomes the fact that he is more open about his emotions, and he has certainly become more expressive about how he feels and what he fears. As we reflect on these changes and his history,

I share some of the data suggesting links between high testosterone, moral decision-making, and empathy. I've had this conversation with many men over the years, and it always seems to elicit a wave of personal revelation.

These days, Aaron is taking time to get to know the people around him. Small talk about shallow subjects can be an easy escape for people unwilling, or unable, to delve deeper, and he tells me that two years ago the extent of many of his conversations would be, "Hey, how about those Giants?"* Now as he greets the staff in our clinic, his salutations go beyond the superficial. Where are they from? Do they have kids? The life stories of others are now of interest to him: people are no longer tools, a means to an end. At most of our appointments these days, I have to give updates on *my* family before we get to him! Outcomes like this are common for patients undergoing hormonal therapy. As their testosterone drops, I find my patients becoming better listeners, gentler, and more considerate. Their spouses often tell me they become better husbands.

In spite of these benefits, my research seeks to treat prostate cancer without the need for hormonal therapy, and I hope someday we can develop new therapies that won't require any testosterone manipulation at all. That said, I think that in the cosmic order of things, the fact that reducing testosterone in these aging men may lead to increased empathy, more emotional engagement in relationships, and a softening of aggression could be something of a silver lining.

Like many patients, Aaron regards these developments with a measure of surprise. Hormonal therapy hasn't been as bad as he expected, and he admits he has actually come to appreciate some of the effects it has had on him. Aaron is fortunate; we'll be able to stop his hormonal

---

*Steve Almond of the *New York Times* commented on the link between sports and virility quite eloquently: "We look to pro sports as a reminder that it is our duty to conceal the parts of ourselves that feel vulnerable, the parts we associate—erroneously, but inextricably—with the feminine."

treatment soon, and odds are that his testosterone will bounce back. We do this in patients like Aaron, who don't have visible tumors but have a rising PSA; we call it intermittent hormonal therapy. For obvious reasons, stopping the hormonal therapy for a bit can improve quality of life. He still fears the cancer, and that's appropriate. I doubt he'd recommend anybody go through hormonal therapy just for the psychological experience of it, but, like many other survivors, this challenge has changed him for the better, leaving him more emotionally in touch, more complex, and, yes, stronger.

Now, a major case is heading to trial and Aaron is the lead attorney. Will having a testosterone level at 10 percent of normal affect his performance? I have to admit I'm a little worried that he may not have the killer instinct in the courtroom that he's used to. I fret and hope he won't blame me if he loses. Oddly, even though I know I am making the right clinical decisions, I sometimes feel a touch of guilt using hormonal therapy on patients—especially on men like Aaron, whose virility is so clearly important to his self-worth. How will he fare without that tool in his back pocket? How important has testosterone been to his lifetime of success? Will he fail without the familiar rush of hormones urging him to take risks in hopes of winning big?

Or, just maybe, will he be better than ever? Perhaps his increased capacity for empathy will allow him to formulate a more emotional argument, or touch a juror in a way that might not have been possible before. Will compassion pick up where competition left off?

Chapter Two

OF MOLECULE AND MIND: HOW FETAL TESTOSTERONE SHAPES OUR THINKING

A lex came into the world healthy and with no complications. His mother had a routine labor and delivery in a well-staffed downtown Chicago hospital, and there were no known genetic diseases waiting to strike from the sidelines. Neither parent smoked, or drank to excess. There was no silly controversy over vaccines. It was 1983 and Alex was, by all accounts, a typical baby.

Yet, by his first birthday, it was obvious that Alex was not picking up on the visual and verbal cues of the people around him. While other kids his age were babbling and actively interacting with their surroundings, Alex remained quiet, almost insular. Cuddling, affection, hugs, and kisses, although lavishly offered to him, were met with little or no reciprocal affection. He showed no signs of talking, and when his parents tried playing peek-a-boo, all that returned was a blank stare. Regular checkups showed that Alex was physically healthy, but by the time he turned three there was still hardly a glimmer of vocabulary. He used a few nouns here and there, but no personal expressions—not even the toddler favorite "I want"—and no third-person phrases like "Daddy want" either. There were certainly no statements about how he was feeling. Alex's verbal communication simply wasn't developing, nor was his use or understanding of nonverbal cues, like facial expressions and body language.

The diagnosis was autism.

With his language skills still severely limited at age four, Alex's parents went in search of a more comprehensive approach to communication—one that would allow their son to finally express his thoughts in some way. Alex's mother, Susan, devoted her full time and energy to bringing in the brightest available minds to evaluate Alex and get things moving.

Alex is now in his midthirties, and I've known of him since the tender age of four because one of the bright minds brought on to his case was my sister Martha. She was a young speech pathologist, fresh off her master's training at Northwestern and working at a Chicago school devoted to autistic education. What turned out to be a very rewarding freelance gig for Martha also helped, in a way, to bring Alex into the world a second time.*

AUTISM, EMPATHY, AND TESTOSTERONE

Most of us possess the ability to perceive and reflect on the mental states of others—we understand that other people have thoughts, feelings, desires, and intentions like our own, and we use this understanding to

---

*Martha's work with Alex also became entwined with our family story, in part because of the poignant pencil sketches Alex's mother, an artist, did of Martha during therapy sessions—sketches that hang in Martha's home to this day.

infer what others might be feeling or thinking. This is known as *theory of mind* or *cognitive empathy*, and an impairment in this area is one of the core features of autism. Someone who lacks theory of mind is said to be "mindblind"; the medical term is *alexithymia*, which, broken down into its Greek components, means "no" (a) "words for mood" (lexithymia). This characteristic is reflected in the difficulty an autistic person experiences in social interactions. As you can imagine, recognizing that others have feelings and being able to imagine what those feelings might be is key to empathizing with them.

Given what we know about how testosterone affects empathy, we can begin to understand a theory developed over the past decade and a half of autism research, one known as the "extreme male brain" (EMB) theory.[14] This theory is not without controversy, but from the perspective of relating testosterone to behavior, it's at least a convenient model for us to examine. In short, it builds on the empathizing-systemizing (E-S) theory, which says that female brains are biologically programmed to empathize while male brains are programmed to "systemize" (literally, to be interested in the analysis and/or construction of logical systems); it proposes that people with autism-spectrum disorders thus have "extreme male brains." Children diagnosed with autism typically share a group of behavioral traits that includes impaired empathy, infrequent eye contact, avoidance of group interactions, and late language development, plus a whole host of other traits more common in boys—even otherwise neurotypical ones—than in girls, ranging from obsession with games and systems to short attention spans to an explosive response to being reprimanded. Boys are three times more likely than girls to be diagnosed with moderate or severe autism, and Asperger syndrome, a milder autism-spectrum disorder, is ten times more likely to be diagnosed in boys than in girls. Brain size, while not related to intelligence, *is* correlated with sex—boys in general have larger brains than girls, and autistic boys have even larger brains than do their non-autistic male peers.[15]

Of course, while autism-spectrum disorders are often discussed in terms of their negative manifestations, ASDs are also associated with a host of positive qualities, many of which are considered assets in modern society, such as attention to detail, or deep passion for a particular area of knowledge that enables an individual to excel in that arena. Autism exists on a spectrum of severity, and many of its key characteristics can be seen in all of us to varying degrees. We'll talk more about this later—for now, let's look at the data suggesting that the process that determines where each of us lands on this spectrum may begin in utero, with testosterone.

Kids and adults with autism don't have higher serum-testosterone levels,[16] but it appears that their brains may have been exposed to higher levels of testosterone before birth. Although high levels of testosterone exposure in the womb do not necessarily result in a child's being born with autism, studies have shown it to be a contributing factor, and these levels can also be associated with characteristics broadly attributed to autism and related disorders, including, among others, low levels of empathy (as we explored with the trolley problem in the previous chapter), as well as differences from the general population in the areas of eye contact, vocabulary/verbalization, and obsessions known as *restricted interests*, all of which will be discussed in this chapter.

In Chapter 1, we discussed the 2D:4D ratio as a marker of fetal testosterone exposure. Recent research shows that the amount of testosterone in a woman's amniotic fluid may be another (if less convenient) way to quantify the amount of testosterone that stimulates a fetus's brain.

So where does this testosterone in the amniotic fluid come from, and how does it affect the developing fetus? It might come as a surprise that much of it is made not by the mother or even the placenta (which handles the bulk of hormone production during gestation) but by the fetus itself. Fetal production of testosterone begins in about the thirteenth week of gestation and spikes very high—almost to the same levels found in an adult—at about week fifteen after conception, after which it drops back down until puberty. As this hormonal flash occurs, the brain is in maximum growth mode, and the areas that grow the most during this time of testosterone stimulation are those governing many actions, behaviors, and traits relevant to autism.

Psychologist Simon Baron-Cohen and his team of researchers at Cambridge University pioneered research into testosterone and autism in the early 2000s. To test their hypothesis that fetal testosterone levels could be related to the behavioral aspects of autism, they set out to observe behavioral characteristics of young children whose prenatal testosterone exposure could be measured with the help of the "amniotic fluid bank" at Addenbrooke's Hospital in East Anglia. The fluid came from the amniocenteses of several thousand women who had undergone the procedure to test for possible pregnancy complications and, after the testing was done, had consented* to have the fluid stored for future research. That's where Baron-Cohen came in.[17]

Baron-Cohen's researchers contacted a large number of the mothers whose amniotic fluid was available for testing, and asked them to bring their young children in for evaluation. The team divided the children into groups by age—under one year, eighteen to twenty-four months, and close to four years old—and studied their various behavioral traits. None of the children had been diagnosed with autism.

---

*"Banked" tissue, like that described here, is an invaluable resource in medical research, despite the fact that it rarely leads to a benefit for the tissue donor him/herself. Despite the recent controversy and appropriate conversations surrounding abuse of such research, many advances in our understanding of disease and development of new treatments come from tissue donated by patients and individuals who have given their informed consent.

FACES, EYES, AND MINDS

One aspect central to the study was evaluating what they called the *state of mind* of each child. Now, if you've ever been around an infant, you know that finding the answer to that question is not as simple as asking the baby to share his or her inner thoughts. Fortunately, a number of methods exist for evaluating the inner life of young children; among other things, this study evaluated each child's tendency to look at faces, appropriately termed *face gaze*. Face gaze, which is distinct from the two-person interaction of eye contact, means simply that the child is looking at another person's face, and in young children it can be used to evaluate the ability to connect with others.* (Face gaze and eye contact are both important but evaluated separately.) Face gaze appears to be governed by an area of the brain known as the *amygdala*, a sort of switchboard between various parts of the brain, and some researchers have asserted that aberrant connectivity of the amygdala to other structures in the brain is a root cause of autistic traits, making the autistic child or adult less likely to engage in face gaze considered "normal" for his or her age group and therefore less likely to empathize and connect with others. The amygdala is perhaps the most important brain structure affected by variations in testosterone, as it is the central hub of emotional processing in the brain, and it also displays slight differences in anatomy based on sex: it is typically bigger in males than in females (although there is a range of size within each sex, and some overlap between them).

Along with studying face gaze, the Cambridge team measured the propensity for eye contact in the children approaching their first

---

*Not surprisingly to many canine lovers, dogs have strong face-gaze aptitude, and most will reflexively look at the right side of the human face—the side thought to convey more emotion than the left. Dogs' leftward face gaze is human specific—they don't do it when they look at other dogs or other animals. It is believed this is one sign of their emotional empathy.

birthdays. Eye contact is frequently the starting point for social inter-action and is a reasonable surrogate for it at this early stage. Over an observation period of twenty minutes, the girl subjects made eye con-tact with a parent approximately twenty-two times, while the boys did so only sixteen times. For people who study such things, this was not a surprise. However, the researchers then found that within both the boy and girl subsets, each child's fetal testosterone level was inversely proportional to his or her frequency of eye contact: whether they were boys or girls, children born from women with higher testosterone lev-els in their amniotic fluid made less eye contact than did those from the mothers with lower testosterone levels. As the researchers explored other behavioral traits, they found, fairly consistently, that the babies who displayed fewer sociability traits—such as pretend play, language use, gaze following, and social (back and forth) communication—at about twelve months of age were the children born from the moth-ers with more testosterone in their amniotic fluid. Similarly, when Baron-Cohen's researchers studied the children aged eighteen to twenty-four months, they found that the ones who had been exposed to high levels of prenatal testosterone were, on average, also "less social"—as based on the size of their working vocabularies—than those exposed to lower levels of testosterone. On the whole, the girls had larger vocabularies than the boys, averaging about ninety words to the boys' forty. Continuing this pattern, in the four-year-old sub-jects the data showed that the higher the fetal testosterone, the less likelihood of the children's having "quality social interactions," which was defined as their number of friends, their sociability at play, and other factors that indicated the children were connecting with others. Remember, in each of these age groups, the subjects being observed were not kids with autism, although the results are a clear indication of why the autism spectrum is called a spectrum: traits associated with the disorder are spread throughout the population to varying, usually lesser, degrees.[18-20]

While frequency of eye contact and working vocabularies both go down with higher testosterone, spatial reasoning (that is, the ability to do things like mentally rotate an object or navigate a maze) increases. This relationship supports other hypotheses regarding behavior, including one that says young children who have difficulty empathizing or engaging in social activities may compensate by spending more time engaging in what is called "systemizing." Systemizing is the process by which one constructs systems, sees patterns, and attempts to predict and control outcomes based on these systems and patterns. We perform systemizing behaviors all the time, when we press a button, interpret a subway map, and so on: we know that when we press a certain button on a keyboard, a certain something will happen as a predictable consequence. It is possible that those with deficits in empathy attempt to predict the actions and emotions of others via systemizing instead.

As Alex grew into adolescence, he could remember family friends and neighbors by their license plate numbers, and to this day he can also recall the room number of almost every hotel in which the family has ever stayed. These are both forms of systemizing.

The importance of these findings is what they tell us about prenatal testosterone exposure and a person's ability to connect with others. In the case of eye contact, we're all aware that looking at another's face can help us discern his or her general emotional state—afraid, surprised, nervous, angry, happy, etc. As the saying goes, the eyes are the window to the soul, or, for our purposes, the eyes are a shortcut to the mind, and with our ability to observe a person's emotions comes a chance to understand and empathize with him or her. Discerning the mental state of another person by looking at the eyes is often a problem for those with autism. More broadly, differences in how easily we are able to identify emotions by looking at a person's eyes may be evidence of testosterone's effect on us all, autistic or not.

To examine the extent to which individuals are able to interpret emotions using eye contact, Baron-Cohen's group at Cambridge developed the Reading the Mind in the Eyes Test, or RMET.[21] Basically, a subject is shown a picture of a pair of eyes and then is evaluated on whether or not he or she can identify the mood conveyed by the eyes' expression. The results of the test have given us two relevant findings. First, individuals with Asperger's or high-functioning autism scored substantially lower than neurotypical controls, in a finding consistent with the hypothesis that autism spectrum disorders are associated with reduced emotional perception and cognitive empathy. Second, when the control groups (i.e., those without autism or Asperger's) were graphed, the population showed a near-perfect bell curve in ability to correctly identify emotions based on a person's eyes, and at one end of the curve the control group's data overlapped with that from the test group. In other words, some of the subjects who were not identified as autistic had "mind reading" scores that were very similar to those of the autistic test subjects. What this means is that a decreased ability to discern emotions in this way is more common in, *but not specific to,* people with autism-spectrum disorders. So what does it mean that many people without autism may, in fact, share a trait associated with autism? And might it be tied to our prenatal exposure to testosterone? Studies in which women were given testosterone before taking the RMET support this possibility—after testosterone administration, their cognitive empathy (as expressed by their scores on the RMET) decreased.[22] To find out, we would need to look specifically at the prenatal testosterone levels of people later diagnosed with autism.

I took the RMET* as a forty-seven-year-old adult and take a little pride in scoring 30/36, which I think is pretty good, important for someone who deals with patients in often-complex emotional situations. But how much of that score comes from my fetal brain and how much of it comes from living in the world and interacting with

---

*The test is available for anyone to take online: https://well.blogs.nytimes.com /2013/10/03/well-quiz-the-mind-behind-the-eyes/.

people for most of those forty-seven years? Am I scoring the way I am because of my biology, or have I learned to overcome my biology? Has my interaction with hundreds of cancer patients, not to mention family members, friends, and even total strangers, helped me to "cultivate" the abilities measured by the RMET, or is this my "preprogrammed" value?

What I found especially interesting about taking the test was that while many of the emotions came to me right away, others I really had to think about. Eyes conveying flirtatiousness, desire, and skepticism were among those I got relatively easily—but would I have at age seventeen? If higher testosterone suppresses our ability to intuit these emotional states, and we do get better at it over time, is that because we intellectually figure it out, or is it due to the waning effect of virility as we age? Is it mind, or molecule?

Another interesting detail uncovered during Baron-Cohen's research was that the four-year-olds with high fetal testosterone exposure were more likely to have *restricted interests* compared to the "lower fetal T" kids.* As you might intuit, restricted interests are what we may lightly refer to as "obsessions": dinosaurs, cars, a certain animal, a particular game, etc. A "restricted interest" is not, for example, just really loving baseball, it's being interested in baseball to the exclusion of other topics. For the most part, these fascinations are not a big deal and are dismissed as a phase (which they usually are), but in some cases they persist over longer periods and can become maladaptive—for example, if they impair school performance or cause other problems. Restricted

---

*The term *restricted interests* is a specific domain defined by the Children's Communication Checklist (CCC), a tool developed by the United Kingdom's government-sponsored Medical Research Council and used to assess the categories of childhood communication, ranging from vocabulary to the quality of social interests to the variety of social interests. This test is administered to young schoolchildren to determine whether a student has special needs that merit attention from the school. The behavioral and social domains in Baron-Cohen's analysis rely on those supplied by the CCC.

interests are a common trait among those with autism spectrum disorders, and in many cases this characteristic is a type of systemizing (like knowing the stats of every baseball player in the history of one team). But again, this trait isn't specific to autism alone, and I'd imagine the association becomes murkier with children whose interests are more than usually restricted but not restricted enough to fit the clinical definition. After all, a child's range of interests is likely affected by everything from attention spans to trends in parenting (the alternating manias for encouraging "well-roundedness" and "finding a passion"). I collected Matchbox cars, as did many boys in my generation. I even had a very cool blue plastic carrying case for them. I miss those cars (especially the gull-wing Corvette and the black Trans Am, which were my favorites) and I bet that collection is worth something today—leading me to ponder whether avid collectors have higher prenatal testosterone as well. Linking the kinds of things we are accustomed to thinking of as personality traits to hormones raises as many questions as it answers.

Autism researchers continue to debate the validity of the "extreme male brain" theory, but regardless of where they settle, I am fascinated by what the EMB theory and aggregating research relating testosterone to empathy and autistic traits suggest about the overlaps between biology and behavior, testosterone and thought. And I wonder whether this might ultimately affect our understanding of the concept of the "mind" in a much bigger, philosophical way.

FETAL TESTOSTERONE AND MIND-BODY DUALISM

The interaction of the winner effect and fetal testosterone is at the core of the intersection of biology and behavior—not only does winning increase testosterone, which drives us to repeat the behaviors and increase testosterone further, but some of us also may even be "wired to win."

In the previous chapter, I discussed testosterone's relationship to the winner effect, using findings from Ben Coates's research on London day-traders. His team's experiments started out by measuring current testosterone levels in the subjects' saliva, but later research found that *fetal* testosterone levels were also in play. The data showed that the year-end bonuses (which are tied to the profitability of their trades) of traders with low 2D:4D ratios—an overt sign of high fetal testosterone—were almost ten times greater than those with high 2D:4D ratios. In fact, the low 2D:4D ratio was even more predictive of an individual's success as a day-trader than his years of experience on the job.[23] Fetal testosterone may influence us in ways we haven't yet imagined. Studies by other researchers have tied this ratio to athletic ability, sexual orientation, fertility, libido, and the risk of prostate cancer. If you enjoy the rush of adrenaline that comes from gambling, extreme sports, or high-risk business dealings, there's a good chance your brain was wired to react that way before you were born. But does our wiring determine our destiny?

To say that certain behavioral characteristics or ways of thinking and feeling are strictly a biological function directly challenges one of the cornerstones of Western philosophy. *Dualism*—the idea of a clear and inviolable separation of body and mind—is as old as Plato, but the more modern branch was popularized by the seventeenth-century French philosopher René Descartes. You may have heard the mind or soul referred to as "the ghost in the machine." This line of thinking holds that while the physical body is bound by the laws of nature, no more than the sum of its chemical parts, the mind or soul is eternal and separate, our thoughts and feelings constituting something more than the product of biology.

In the twentieth century, however, the tenets of Cartesian dualism were challenged both by the influx of Eastern philosophy and advances in biological research, especially the study of clinical disorders involving the brain. The more we learn about neuroscience, the more we discover that much of what we think of as the activity of the mind—our thoughts, our moods, our memories—has a physical, chemical basis. While Descartes proposed that "our minds are our 'selves'" and that the mind is not "corruptible" (my word) by the body, other lines of thought, Buddhism among them, proposed that we are more holistic in nature—an enmeshed combination of physical, psychological, and experiential components. So the question arises, if we seek to manipulate the behaviors driven by testosterone, are we manipulating the body or the mind, or both? After all, the brain is plastic; that's how we learn. Non-medication treatments for conditions like autism and ADHD—social-skills groups, for instance, or behavioral therapy—have the capacity to rewire the connections between neurons. We can physically change our brains by changing our behaviors.

There are no easy answers in philosophy, and, in a way, Coates's research on the winner effect, too, tangles with these questions. His data is consistent with the idea that a risk-taking brain—the product of high fetal testosterone—is also likely to have a higher concentration of androgen receptors, one of the parameters that determines how much effect testosterone will have on any given individual. This surge of testosterone that comes from success thus reinforces the behavior and increases the likelihood of even riskier behaviors. Of course, it is not just the number of receptors that determines the effect of testosterone, it is also their sensitivity, and the picture of the effects of fetal testosterone is a complicated one. In the trolley experiment we saw that women with *low* fetal testosterone were more susceptible to testosterone surges. When it came to making a moral decision, testosterone reduced their equivocation and allowed them to accept killing when necessary. When one considers that a surge in testosterone may decrease moral equivocation and increase risk taking, we can see a potential peril of virility, but we can't say that this overrules our human ability to reason—can we? What's the mechanism at work here—the body or the mind? No one can say for sure. What's clear is that biological forces can and do influence our thoughts and behavior, and manipulating the former can have effects on the latter.

The more we understand and accept autism, the better life can be for those living with it, and research over the past few decades has taught us more about the disorder than ever before. Working with my sister and others, Alex learned skills that enabled him to interact with the world and share himself with it more fully. He's also benefited from modern technology, using tools such as Tumblr, FaceTime, and text messaging to help him connect with others in ways that are comfortable for him. When there is a need for an emotional conversation, Alex will sometimes go to another part of the house and will Facetime his mother—the relaxation produced by even that minor distance is lowering the communication barrier.

Still, some are holding out hope for a cure, or for medical, not behavioral, treatments to ease the disorder's negative effects. The cause of autism is extraordinarily complex and can't be reduced to the effect of a single hormone at week fifteen of gestation, and yet the data suggests we should continue to explore the connection. Given what we've learned about prenatal testosterone exposure, could testosterone suppression play a part in future treatments for autism? We don't have the answer to that question (yet), but while blocking the effects of testosterone, as I do to treat my patients' prostate cancer, has for the most part not been used to treat autism, administering *oxytocin*—the hormone that counterbalances some of testosterone's effects on the brain—is showing promising results.

As part of my work as the chair of one of UCSF's Investigational Review Boards (IRBs), whose job it is to review research proposals and ensure they pass academic, scientific, and ethical muster, I think a lot about the ramifications of medical experiments on their human subjects. One series of studies the committee reviewed and eventually approved a couple of years ago tested the effects of oxytocin on autistic children, although not in the way you might think. The hormone, which has been shown to increase "maternal behavior," including empathy and bonding, was administered intranasally *to the mothers* of young children with autism, the hypothesis being that the higher levels of oxytocin in the mother will increase her affection for and sense of bonding with the child, a psychological change that might in turn encourage eye contact and other behaviors affected by the child's autism. I was struck by the way in which this series of experiments underscores our biological connectedness; it had a certain beauty to it.

Oxytocin is, in many ways, a biological foil for the effects of testosterone on behavior. In mothers who have just given birth, oxytocin surges while testosterone plummets, and studies have shown that whereas testosterone will suppress empathy, oxytocin will directly promote it, in both men and women. In an experimental setting, women with high testosterone have been shown to pay less attention to the face of a baby than do low-testosterone women, but when the high-testosterone women are given oxytocin, they respond like the low-testosterone women. There are variations in individual reactions to oxytocin, dependent on the number and sensitivity of oxytocin receptors, just as the effects of testosterone depend on androgen receptors.[24]

In the last couple of years, the first studies of oxytocin being administered to autistic individuals themselves suggest that using the hormone every day for five weeks may have some benefits. During that time, the autistic test subjects became more sociable, and their scores on the Social Responsiveness Scale (SRS)—a questionnaire (filled out by parents) about eye contact, language, and other facets of human interactions—improved.[25] Longer-term studies of oxytocin use in autistic children are currently underway.*

---

*Intranasal oxytocin administration is also being tested for its ability to lessen suffering from post-traumatic stress disorder (PTSD) and even to help recovering alcoholics.

We know that oxytocin promotes bonding and caring behaviors, and that these behaviors, in turn, can actually cause oxytocin levels to rise, creating the same sort of feed-forward loop we saw in the winner effect. What we don't know is *how* plastic this system is. If we teach a high-testosterone/low-oxytocin mother to express traits associated with higher levels of oxytocin, will the feed-forward loop kick in and be self-sustaining? Maybe, maybe not. But at the very least, these studies show that we may be able to overcome some of our hardwiring; it's a finding that I believe has important implications for human behavior—and even human nature—as a whole.

I hold dearly to the notion that a theory of mind is innately, and uniquely, human, and also to the idea that it can develop over time, as I believe it has for Alex. As a teenager he was mainstreamed into the school district of a rural exurb of Chicago, and then he attended community college. Now in his thirties, he is continuing to thrive, living a full life surrounded by people who treat his autism not as a disease to be cured but as a difference in perspective, one that merits understanding and respect. Alex is known for his photography, artistic talent, and general good nature. His mother shares with me in an email that every half hour or so over the course of the evening, as they sit around together, Alex will ask, "Is there anything you need, or want?" This is not a question that would come naturally to him, but he has learned how to perform the give-and-take of relationships. He may have been born hardwired with what could be considered a "deficiency" in this area, yet his growth is proof to me that the capacity of the mind is not bound by the biology of the body.

Chapter Three

WOMEN AND TESTOSTERONE: INNER VIRILITY AND OUTER IDENTITY

I n terms of physical appearance, the embryos of both male and female humans remain undifferentiated until about the seventh week of gestation, at which point the developing body—and the genitals in particular—become either outwardly male or female. This is how it works: although sex differentiation is determined by chromosomes (either XX or XY) at the time of conception, the initial trajectory of every developing embryo is female. This means that unless some processes are turned off and others turned on by the presence of the Y chromosome, all embryos will develop female anatomy. The introduction of testosterone (and enzymes that drive this and other systems) during gestation is one of the ways that trajectory is altered, causing the embryo's development to turn off the female path and onto the male path. There is no spike in estrogen levels that turns the embryo into a female, only a surge in testosterone that turns it into a male. One can play with the notion that there are cosmic philosophical meanings associated with the idea that "female is the default pathway," but there are also real, timely, and pertinent questions to be explored regarding what it means to be a woman, both physically and psychologically, and what role the hormone testosterone plays in women in general.

THE FEMALE TESTOSTERONE CURVE

Despite its reputation as a "male" hormone, testosterone is a normal part of the female hormonal makeup, and its effects are not confined to outliers with either too much or too little of the hormone. As a group, women do have less testosterone than men, but that's not the whole story.

Serum-testosterone levels in women are about 5 to 10 percent that of their age-matched male counterparts, although large amounts of other androgens can be found in the female system, including DHEAS (dehydroepiandrosterone sulfate), DHEA (dehydroepiandrosterone), and androstenedione (these other hormones are chemically very similar to testosterone and have the same function but are weaker). Whereas in men more than 90 percent of the testosterone is made in the testicles, in women it comes from diverse sources: about 25 percent is made in the ovaries, 25 percent in the adrenal glands, and the rest from the conversion of the other steroids (mostly androstenedione) that circulate in the blood. As with men, women experience a gradual decline in serum-testosterone levels beginning around age thirty.

Surprisingly, menopause, during which estrogen levels drop significantly, has little or no effect on testosterone production, and as estrogen levels fall and testosterone levels remain stable in menopausal women, the *relative* effect of testosterone may increase. While it is true that men and women have *quantitatively* different levels of testosterone, it would be an oversimplification to say that's what makes men and women different, and indeed women *qualitatively* experience many of the same effects from testosterone that men do. Everything we discuss in this book as an effect of testosterone, from boosting confidence to fueling competition, from increasing libido to building muscle, applies in the female body as well as the male one—it is simply a matter of degree. From behavior to baldness, empathy to aggression, testosterone does what it does to *all* of us. Sugar is sugar regardless of what recipe you put it in, but both the amount you add and what you add it to determine the end result.

One of the important subtleties to recognize when talking about differences in testosterone in men and women is that testosterone levels are of course only one piece of what produces testosterone's effects. For instance, while testosterone contributes to libido in both men and women, there is no evidence to suggest that a man with normal testosterone levels for a male experiences a much greater libido effect than a woman with normal testosterone levels for a female, even though those levels are likely hundreds of ng/dL apart. Rather, women are sensitive to much smaller quantities of the hormone, and a testosterone level high enough to cause unwelcome symptoms or side effects in a woman would not even crack the low end of the normal range in a male. The effects are similar, but the range at which these effects present themselves are not. However, the fact that women have quantitatively less testosterone and more estrogen than men does lead to qualitative differences—some obvious, like muscle growth, facial hair, and so on, and some less so.

For instance, men have a reputation for being oblivious. I know I do. It's not uncommon for my wife to remind me several times about an upcoming event, or to repeat information several times before I internalize it. I know I'm not the only husband asked regularly, "Do you hear anything I say?!" I would *like* to attribute this slight flaw in my communication skills to the fact that I am, well, a deep thinker, always pondering something bigger than the questions of whether the dog has been fed or whether I have my kid's upcoming school play on my calendar. Alas, the truth is I *am* oblivious from time to time.

The good news is that science has my back—you might even say I can't help being this way. Research has shown that higher fetal testosterone levels allow the brain (or perhaps "cause the brain" is more generous) to filter out what it considers "noise." Studies have shown that whereas the sound-processing area of women's brains became active in response to hearing both music and random noise, men's brains generally responded to the music only.[26] In a way, this "noise canceling" is a hardwired part of male hearing.

Not surprisingly, higher levels of fetal testosterone in females are associated with what scientists call a "male pattern of auditory recognition," which means, in short, that the higher the levels of prenatal testosterone they were exposed to, the more likely they are to filter out what they interpret as "noise." Here again, I think calling it a "male pattern" is misleading and inaccurate; auditory recognition varies across a range, and although that range is affected by fetal testosterone, it's not a strictly male/female issue. That said, men are *generally* more likely to have been exposed to higher levels of prenatal testosterone and are therefore *generally* more likely to fit this pattern of auditory processing.

How does this work on an individual level, and, more specifically, in my own marriage? It is actually quite amazing to watch my wife juggle everything she has going on. She works full-time as a librarian and manages to pull off several hundred library programs each year, and at home, she pays every bill and maintains all our household files. Although we share the cooking (I do 30 percent—OK, maybe less), she's the one who remembers to shop for not only dinner but also the kids' lunches and household supplies like toilet paper and dog food. When I shop, I'm typically hungry and focused only on my next meal. I do think this ability to focus on a single subject is one of my strengths (in the right context), and although it sometimes annoys my wife, whose strength is to think broadly, to see the world through a wide-angle lens, I've noticed that a lot of heterosexual couples are this way—and that's likely not an accident. There are cultural and sociological factors at work, of course, but it is likely that evolution, and the different kinds of focus required by childrearing and, say, provisioning, have something to do with it as well.

THE RISE AND FALL OF HORMONE SUPPLEMENTATION IN WOMEN

Before we delve into the world of female testosterone supplementation, let's establish the difference between the *endogenous* and *exogenous* forms of the hormone. The former, from the Greek meaning "coming from inside," describes testosterone produced within the body itself, while the latter ("coming from outside") refers to testosterone administered as a treatment.

When you hear about people taking supplemental (exogenous) testosterone, you probably think of men, and whether or not the treatment actually provides all the benefits it promises, the testosterone-supplementation industry is doing booming business, making many billions of dollars each year. Less well known, however, is the fact that women have also jumped onto the bandwagon, and often for similar reasons.

Back in the 1990s, *hormone replacement therapy* (HRT), particularly for menopausal women, was in its heyday, and many women (and their doctors) were experimenting with the idea that vitality could be preserved and even enhanced through the administration of supplemental hormones, including estrogen and *DHEA* (dehydroepiandrosterone), a steroid used to increase testosterone levels.

When HRT was popular, psychiatrist Louanne Brizendine, a colleague of mine at UCSF and the author of both *The Female Brain* and *The Male Brain*, incorporated DHEA supplementation into her clinical practice and was able to observe and measure the satisfaction of the women who took it. She describes its effect as a trend moving from "feeling" to "focusing," and she found that in women the increased focus on fewer things at the expense of being more widely aware became, ironically, a distraction. "It did a lot for their libidos," she told me over lunch near our campus. "But, for many of them, becoming oblivious to their surroundings made them uncomfortable." The most interesting question for me is not why the DHEA led to intense focus and a lesser ability to think and act more broadly, but why this state of mind was so *dissatisfying* for her patients. Was the problem really the change itself or simply the fact that it was a change at all, especially one coming later in life? Perhaps one person's focus is another's obliviousness? Given the balance in my marriage and family life, such a change may lead to unfed kids, and fewer programs at the library. Who knows?

Unfortunately, the optimism surrounding HRT came crashing down when the data from a large international randomized trial revealed that HRT increased the risk of cancer and negative cardiovascular side effects, and as a result of those findings the practice of HRT is no longer widespread.[27, 28]

TESTOSTERONE SUPPLEMENTATION TODAY

While testosterone supplementation for women hasn't gone away completely, it's now done more or less under the radar. DHEA in particular continues to be a big seller in the nutriceutical market and is still available over the counter in pill form. What we don't know is *how much* DHEA sold over the counter is actually being taken by women, and what its effects are. As a result of the Dietary Supplement Health and

Education Act of 1994, the sale of DHEA and similar hormone supplements is not regulated by the FDA, and the efficacy and safety of such substances does not need to be guaranteed by the manufacturer. At low doses most women will experience improvements in libido and mood, and DHEA could help with losing fat and gaining muscle, but as doses rise there are also side effects many women would find to be undesirable, including characteristics typical of anabolic-steroid supplementation, such as the development of bulky muscles, increased growth of facial hair, a lower voice, and so on.

The writer Ann Mallen once spent several weeks figuratively walking around in a man's shoes due to a pharmacy mix-up that had her dutifully applying testosterone cream to her skin every day. As a result, her appetite for sex soared and became a "constant distraction," and as her fuse shortened she found herself wracked with bouts of what she called "irrational anger." The error was discovered a month later and she resumed her normal life, but for Mallen, this was a brief but life-changing peek through a window into how the opposite sex thinks and feels, and in her article about the experience for the *Washington Post,* she concluded that "but for a simple hormone, it is possible to live as either a male or female." This is an oversimplification, but it goes to show how a thin hormonal veneer may separate what we think of as "femaleness" from "maleness." "Underneath the high-pitched whine of our sex hormones," Mallen wrote, "we are neither."[29]

Unlike Mallen, a woman I know named Karen uses testosterone cream on purpose, and has no intention of giving it up. Karen is an artist, skilled in mosaics, and her spacious Minnesota studio is filled with natural light and packed with hundreds of jars of Italian glass in every shade, imported from around the world. The color and the glass are her antidote to the long Midwestern winters, and it's in that season she does her best work.

Although she doesn't describe it as depression, Karen says that before starting hormone supplements she was struggling with a sluggish libido and overall low energy, as well as vaginal dryness. "At sixty," she says, "I was too young to feel old, and I wasn't always engaged with life the way I wanted to be." Her primary physician suggested a bit of testosterone cream to see if it could brighten things up.

It was a natural choice for Karen, who was inspired by the positive experience her husband had while taking testosterone supplements himself. He's fifty-two and has been on testosterone pellets for about fourteen years.

"My husband and I haven't really had any issues with sex, yet I didn't always want to make the effort. I'm going to guess that my testosterone was never that high," she said.

About eight weeks into the supplementation plan at the time we spoke, she described her experience as you might expect an artist to: in the context of color.

"You know the color wheel—how every color consists of hue and value?" she asked me.

I didn't, so she explained. Colorists describe the properties of color on the basis of its *hue*—the presence of one of the primary colors (red, yellow, or blue)—and its *value*, or the level of lightness or darkness associated with the color. Imagine a spectrum that is pure red on one end, gray on the other, with gradations in between, and you get the idea. I think of it as the "richness" of the color.

"Well, the layers of gray are coming away from me," she said. "It makes me not want to hide from the day. I'm present in our relationship."

Her enthusiasm is spreading with the speed of a book recommendation, and her four best friends are also now on testosterone cream. Decreased libido is one of the most common complaints in women with low testosterone—a common symptom of menopause—and today the condition (now referred to in a clinical setting as "hypoactive sexual desire") can be treated with the help of several pharmaceutical options. While some women take testosterone and/or DHEA, a new drug called flibanserin treats the problem without altering hormone levels but instead acts more like an antidepressant, by acting on the brain's neurotransmitters, not hormone levels. It hasn't turned out to be that popular, however, primarily because it can cause dizziness and can't be taken with alcohol. I suspect a glass of wine and a tiny rub of testosterone cream might be a lot more fun and effective.

In the wake of the HRT boom, we have gotten a glimpse of how the wonders of modern pharmaceutical technology could change the human experience as we know it. In one study, women taking DHEA showed improved visuospatial performance,[30] an ability that might be helpful for, say, pilots, combat soldiers, surgeons, and architects. Imagine custom-tailored hormones that allow us to be exactly what we want to be. I came across a term in a journal that initially made me cringe: "cosmetic neurology"—basically, using medications to improve brain function rather than only to treat medical conditions. Perhaps this where society is heading.

What we still don't know is why, if testosterone works so well for Karen and her friends, it was so unpleasant for Ann Mallen. The answer is probably based on the extent to which an individual's brain has been "primed" to react to the hormone. Once again we are back to examining how testosterone levels affect the fetal brain.

FETAL TESTOSTERONE EXPOSURE IN WOMEN—SENSITIVITY AND SEXUAL ORIENTATION

I think of the fetal brain like molten metal, in that it's malleable and can change a lot with only a tiny amount of pressure. After birth and throughout life there is still some malleability in the brain, but—at least as far as the influence of testosterone goes—that malleability is greatly reduced. One recurring theme in experiments that measure the effects of testosterone on women is that women who had *low* fetal exposure to testosterone (as based on their 2D:4D ratio) are generally more susceptible to the behavioral effects of exogenous testosterone—that is, testosterone supplements. Meanwhile, those who had higher levels of testosterone exposure in utero tend to be less swayed by the hormone when it is administered exogenously. What this means is that the greater the testosterone exposure in utero, the less effect exogenous testosterone will have. The tank, so to speak, is already full. Perhaps Ann Mallen had lower fetal testosterone exposure than did Karen, and so the effect of supplemental testosterone later in life was greater—and more unpleasant—for her.

One of the most fascinating areas of study into the effect of fetal testosterone on women has to do with the hormone's effect on sexual orientation. Studies have shown that self-described lesbians are likely to have lower 2D:4D ratios (that is, a greater difference between the two, indicating higher fetal testosterone levels) than women who identify as heterosexual. Interestingly, no such relationship has been found between gay and straight men. In other words, fetal testosterone levels in men do not appear to have any relationship to sexual orientation.[31, 32] As for hormone levels after we're born, research hasn't yet given us a full understanding of how (or whether) hormone levels in adulthood have anything to do with sexual orientation. (That is, you can't argue that giving testosterone to women will turn them into lesbians.) The science behind sexual orientation has exploded in the past couple of decades, and there are now several genes that are implicated. Taken as a whole, homosexuality in either gender is much more complex than we currently understand. However, at least in women, evidence suggests higher levels of prenatal testosterone have something to do with it, and there may be other links between the hormone and sexual orientation as well, as suggested by some interesting research into a common condition affecting women with excess testosterone.

Polycystic ovary syndrome (PCOS) is the most common cause of female infertility, and is present in up to 15 percent of reproductive-age women. As the name implies, PCOS is a condition in which the ovaries develop cysts, in this case because of hormonal imbalances that interfere with ovulation. The normal range of testosterone in the blood of an adult woman is about 20 to 60 nanograms per deciliter; PCOS women have concentrations that range from 29 to 150 ng/dL. In addition to compromised fertility, other manifestations of PCOS include facial hair, acne, menstrual irregularities, weight gain, and an increased risk of cardiac events.

One study done in London and published in 2004 compared both the testosterone concentrations and the ovaries of women self-identified as lesbians to those of women who identified as heterosexual. Interestingly, 80 percent of the lesbians had cystic ovaries, whereas only 32 percent of the heterosexual women did.[33] In exploring possible connections between testosterone, sexual orientation, and PCOS, research has shown that lesbian women who had non-cystic ovaries did *not* have androgen levels that exceeded the range found in heterosexual women; lesbian women with cystic ovaries, however, had androgen levels that were much higher than did heterosexual women with cystic ovaries. In these studies, the difference was less in testosterone than in *androstenedione* (a precursor chemical that is converted to testosterone and shares many of the same properties). Further, studies of female-to-male transgender (FMT) individuals have revealed that they were about ten times more likely to have PCOS than control women, and more than 80 percent of FMT individuals had elevated levels of testosterone, androstenedione, or another testosterone precursor, DHEAS.[34]

Of course, human sexuality exists along a spectrum, and just as there are high- and low-testosterone heterosexual women, there are high- and low-testosterone lesbian women. It is interesting to note that those who self-identify as "butch," or more masculine, do tend to have higher levels of testosterone compared with those who self-identify as "femme,"[31] and yet it is important to remember that association is not causation. And again, serum-testosterone levels are just one-third of the triad (along with fetal testosterone exposure and the reactivity of an individual's androgen receptors) that determines testosterone's effects.

OUR HORMONES, OURSELVES

Women may not be driven by testosterone to the extent that men are, but they are driven by it just the same, and it plays an essential role in their hormonal makeup, just as it does for men. It seems men and women are also alike in having a topsy-turvy relationship with this particular hormone. For some women, testosterone supplementation can lead to rage, decreased empathy, and even male-pattern baldness, while others experience only the benefits of the hormone, from the bedroom to the boardroom and beyond. For what it's worth, actress/icon Jane Fonda, now in her eighties, attributes some of her ongoing sexual vigor to supplemental testosterone.

Also worth mentioning is the fact that many women manipulate their testosterone levels without even realizing it. More than 10 million women across the country use oral contraceptives each year, and some forms of the pill consistently decrease testosterone levels by 15 to 20 percent. Ironically, the decline in testosterone in women on oral contraceptives can lower the libido, as the hormone binds to a protein known as *SHBG* (serum hormone binding globulin) and therefore becomes inactive. How much testosterone decreases in response to an oral contraceptive is dependent on, you guessed it, the reactivity of an individual's androgen receptors. The more active androgen receptors a woman has, the more likely her libido will be preserved.[35]

(Interestingly, evolution has also developed a mechanism through which women unconsciously or subconsciously manipulate testosterone levels in *men*: studies have shown that when men see women cry, their testosterone levels plummet.[36])

I consider testosterone's effect on women to take a three-dimensional form—there is an effect on behavior, an effect on appearance, and an effect on identity. These are all at work naturally in every woman, every day. When it comes to outliers on the testosterone spectrum, or those who take testosterone supplements, the effects on behavior can be rapid and transient; the effects on appearance are more, shall we say, sub-acute (taking longer to develop), while the effect on identity remains something of a mystery—one we'll consider further in the next chapter.

Chapter Four

BELL CURVES AND BINARIES: WOMANHOOD AND TESTOSTERONE AMONG THE OUTLIERS

Amalia was twenty-four when she first got pregnant. She worked two jobs and was unmarried, the father nowhere to be found. After an ultrasound at her community clinic, the doctor told her she was carrying a boy. Following an uncomplicated delivery at a hospital in California's Central Valley, the obstetrician announced with an unusual pause, "Your baby's . . . beautiful." The doctor usually announced the baby's sex, but holding this newborn, he was unable to confidently do so because the child possessed what appeared to be both male and female genitalia. There was a partially formed penis, but no testicles, and a rudimentary vagina. These details hadn't shown up on the clinic ultrasound.

Amalia was referred to UCSF, where a team of experts take on these cases. Part of the process is chromosome analysis, which in this case showed the baby had neither the XY chromosome pair of a boy nor the XX pair of a girl but an XXY. The next issue was that because the baby's sex didn't fit into the typical male/female dichotomy—that is, since sex was not determined by biology—it was going to have to be a conscious decision. Multiple specialists (urologists, endocrinologists, psychiatrists, and social workers) weighed in, and the analyses showed that the brain of Amalia's child had most likely been bathed in high levels of testosterone during fetal development and would therefore be more likely to develop as a "male brain." Amalia gave her baby the androgynous name Jamie and decided to raise him as a boy. She and the doctors were playing the odds that Jamie would identify as male, but no one could be certain.

The accuracy of their educated guess might not be known for decades, and while the clinical standard is to determine the "success" of such a decision by whether or not the individual grows up to identify with his or her assigned gender role—in Jamie's case as a man—in truth it comes down to whether or not the person feels *content* in the gender to which he or she has been assigned. That said, even people born without chromosomal abnormalities can have gender identities that do not match their genitalia. One high-profile example is Caitlyn Jenner, the Olympian formerly known as Bruce Jenner, who says she experienced gender dysphoria for most of her life before publicly coming out as a transgender woman in 2015.

Intersex and transgender individuals have always been part of our communities, but only in recent years have they become more visible and more widely accepted in society. As visibility increases and attitudes change, so do the ways in which we think about, talk about, and categorize not just intersex and transgender individuals but humans in general. We are no longer confined to the binary categories of man and woman; we see that gender, and even sex, exist on a spectrum.

SEX VERSUS GENDER

When placed on the timeline of human existence, our ability to separate the categories of sex and gender is only just beginning. Only in the last seventy years or so have Western intellectuals and scientists sought a deeper understanding of gender identity, one that breaks from the rigidity of the traditional male/female binary. At the heart of the issue is the "debate" between biological determinism and social constructionism, and that is where the importance of the terms *sex* and *gender* come in. To put it succinctly: female sex is based on a person's biologically determined anatomy, and female gender is a societal construct. What we'll call "woman-ness" is tied in with both.

A towering figure in the development of gender identity and arguably the most foundational thinker in feminism is the French writer Simone de Beauvoir. The most famous line in her 1949 book *The Second Sex* is, "One is not born, but rather becomes, a woman."[37] This statement puts the concept of womanhood in the domain of *phenomenology*—a philosophical school of thought that developed in the first half of the twentieth century and that, at its core, encourages us not to make assumptions but to let our beliefs be validated by actual experience. As applied to sex and gender, the thinking is that simply being born biologically female is not what makes someone a woman. Rather, to call oneself a woman is to rely on, join in, and be defined by the *shared experience of womanhood*.

Seven decades later this influence persists, and modern thinking about gender identity is less focused on the genitals than ever before. Contemporary philosophers have defined womanhood as a "cluster concept," meaning there are discrete parts but that the whole is greater than the sum of these parts, so to speak. Writing in 2000, philosopher Natalie Stoljar, today at McGill University in Montreal, suggested that, like a nation strengthening its borders against a hostile neighbor, defining what it means to be a woman is an important part of defining how the struggle for equality should proceed. She defined four basic categories of the cluster concept of womanhood: (1) female sex organs; (2) shared experiences of womanhood, which include menstruation, childbirth, breastfeeding, and even non-biological aspects such as the experience of oppression or objectification by men; (3) adhering to "typical" gender roles such as dress, behavioral, and work norms; and, finally, (4) gender attribution—that is, calling oneself a woman.[38]

A person can identify with all of these or only some, but essentially Stoljar's theory is that a person who matches this description is a woman. Some have argued that possessing three of the four criteria is enough, and therefore, for instance, transgender women are included. The idea remains up for debate (this is philosophy, after all), but the general idea is that the closer one identifies with what Stoljar has called "the woman paradigm," the more appropriate it is to call oneself a woman. One major problem I see is that this idea opens the door for statements such as "She's not a *real* woman," which suggests that a person who fits *all* criteria is somehow more of a woman than one who does not.

In light of our growing understanding of sex and gender identity, non-binary gender language is emerging more often in public discourse. In 2017, Brown University's acceptance letters went out to prospective students without use of "he" or "she" but instead the gender-neutral singular "they," often to the confusion of recipients. The Ivy League school is among those leading the pack in rolling out gender-neutral language and facilities. Other schools, businesses, and institutions are embracing similar standards, and in late 2016 NBC reported on what is believed to be the country's first non-binary birth certificate. Sara Kelly Keenan of New York was granted a new birth certificate that labels her gender [*sic*] as "intersex" rather than "male" or "female." Keenan was fifty-five years old at the time.

Things are changing, but in some cases it doesn't happen easily or without a fight. One area that has been hotly contested is the role of sex and gender when it comes to who is allowed to use which public bathrooms. Famously, in 2016 the legislature of North Carolina managed to step in it pretty badly, requiring people to use the bathroom corresponding to the sexual identity written on their birth certificates. The legislation was essentially the result of government officials going out of their way to marginalize transgender individuals, but of course intersex individuals were also affected. The state quickly became the subject of ridicule, the object of scorn, and the recipient of numerous costly boycotts, including the relocation of the NBA All-Star Game, the cancellation of a Bruce Springsteen concert, and many corporate sanctions. The legislation was signed by Governor Pat McCrory in March 2016, and portions of it were repealed in March 2017. Meanwhile, the debate rages on.

NATURALLY HIGH

So far we've looked at some issues surrounding gender identity and what happens when it is not aligned with traditional expectations. To go even deeper, let's examine the challenges that arise when a person's gender, sex, or *hormonal makeup* fall outside expected norms.

A century ago, women with naturally extremely high testosterone were often the objects of fascination and exploitation. They were sometimes displayed as circus freaks—usually called werewolves or bearded ladies—and although a few became wealthy and even internationally famous, it couldn't have been an easy life, or one without complications. Today, many of these women would likely be diagnosed with congenital adrenal hyperplasia (CAH), which arises from a mutation in an adrenal enzyme known as *21-hydroxylase deficiency*. This genetic condition alters the flow of steroid-molecule production and leads to excess androgens, which can manifest in a range of severities, from prominent facial hair to ambiguous genitalia.

I am most familiar with the 21-hydroxylase enzyme from the several years I spent conducting clinical trials in men with prostate cancer.

We did extensive testing in our program of a drug called *abiraterone acetate*, which has since been approved by the FDA and used by several hundred thousand men with advanced prostate cancer to increase chances of survival and reduce the risk of recurrence. The drug works by blocking an enzyme that is close to 21-hydroxylase, with the result that it reduces androgen levels to near zero. This same drug is used with great success in women with CAH. From a clinical perspective it is interesting to note that the dose women need to control their symptoms is only one-tenth of what we give to prostate cancer patients—another reminder that when it comes to androgen levels, everything is relative.[39]

Perhaps the best place to explore the shifting sands of hormones and gender is within the world of high-level competitive athletics, namely the Olympic Games. Governing bodies of sports that hinge on individual participation metrics have recently struggled with the "gender verification tests" they have been conducting for years, intending to guard against, for instance, a man masquerading as a woman in order to compete against other women. This type of cheating is exceedingly rare, but the dragnet process of policing gender has also identified a number of individuals with intersex characteristics who were competing within their self-identified gender groups but had physical (including genital) characteristics that suggested they were of the other, or "opposite," gender.

In 2009, when Caster Semenya, an eighteen-year-old from South Africa, won the 800-meter race by more than two seconds at the World Track and Field Championships, a fellow competitor raised suspicion that Semenya was actually a man. The controversy was stoked by Pierre Weiss, general secretary of the International Association of Athletics Federations (IAAF), who said, "She is a woman, but maybe not 100 percent."[40]

Ouch.

Semenya was barred from competition until her gender was verified. The IAAF confirmed that she indeed was female on the basis of her anatomy. One finding that came out of this investigation, however, was that testosterone levels in her blood were three times the normal level for a female of her age.

In 2011, Semenya began hormone treatment (the identity of the drug has not been released) in order to reduce her testosterone levels to comply with IAAF regulations, and she eventually returned to competition. Although her race times suffered, she managed to win a silver medal at the World Championship in Daegu, South Korea, that same year.

The outlook changed for Semenya when, in 2014, the performance of eighteen-year-old sprinter Dutee Chand, from a remote village in eastern India, raised similar suspicion. Chand was also subjected to gender-verification analysis and found to have an elevated serum-testosterone level. Her total testosterone level was not released, but reports indicated that it fell within the normal range for an adult male. (Women, you'll recall, typically have total testosterone levels measuring about 5 to 10 percent that of their male counterparts.) There is no suspicion that Chand was doing or taking anything to raise her testosterone level—it was widely accepted as her natural level—yet the Sports Authority of India judged against Chand. Their statement read: "The athlete will still be able to compete in the female category in [the] future, if she takes proper medical help and lowers her androgen [testosterone] level to the specified range."

Here we have a slippery slope. What's next? Will sports authorities ask LeBron James to be a little shorter? Or perhaps ask Michael Phelps to have his feet made smaller? They wouldn't, of course, and part of why they wouldn't is that when a male has extraordinary athletic traits, no one suggests he should be competing against a different group of people. The issue here is the perceived threat of the *hyperandrogenic* female and the specific concern that her physical attributes place her more appropriately in competition with men. Is there (or should there be) a line somewhere on the spectrum? If so, who is in charge of defining it, and on what basis?

Chand refused to take testosterone-suppression treatment and instead challenged the ruling, arguing that her "natural genetic gift" should not disqualify her. She cited the fact that no other athlete has been disqualified because of a natural physical attribute.

She was right, of course. The case was taken up by the Court of Arbitration for Sport, in Switzerland, and after hearing both sides of the story (including testimony by Stanford bioethicist Katrina Karkazis, a champion of the rights of intersex individuals), the court sided with Chand and suspended the hyperandrogenism regulation, allowing her, and Semenya and others like them, to compete without hormonal manipulation. The court's decision was based on doubt as to whether hyperandrogenism alone was sufficient to create an unfair advantage, and they were not convinced that high levels of testosterone were more important than good coaching, proper training, or nutrition. Perhaps most important, the decision supported the fact that Chand's high level of testosterone was not the result of cheating or malfeasance; it was simply her natural physiology.

The key here is understanding the bell-curve nature of testosterone levels across people as a whole. What we call the "normal range" is not a rigid set of numbers, defined and bound by nature. Instead, we might say "normal" is what we would expect for an individual based on certain criteria (such as gender and age) as compared to the levels of similar individuals. In the lab at my hospital, for example, the normal testosterone range in males is between 240 and 871 nanograms per deciliter of blood. Within that range (more or less) the population follows a bell curve, which is to say most of the adult population is in the middle of this range, while a minority have levels at its upper and lower ends. In females, the normal testosterone range for an adult woman is between 20 and 60 ng/dL, also distributed along a bell curve. Comparing the high end of the female range to the low end of the male range, you can see the vast unoccupied space between 61 and 239 ng/dL; the levels are not even close. This gap is the kind of a no-man's land (couldn't resist the pun) in which we find the true outliers.

When you get right down to it, even putting hormonal ranges aside, athletes who compete at the Olympic level are already outliers. Investigating the component parts of their exceptional physiology, and then mandating that it be stripped away or manipulated, is far more than regulating sport; it is a violation of personal dignity and human rights. Not all that is fair is equal.

LESSONS FROM DOPING IN DEUTSCHLAND

These recent events are not the first time that female Olympians have been caught in a whirl of testosterone-related confusion and controversy. Track-and-field star Heidi Krieger's life was turned upside down in the 1970s and '80s, starting at age fourteen, when she was enrolled in the Sports School for Children and Youth in her native East Berlin, an institution famed for its intense coaching and fast track to Olympic glory. Training was all-encompassing and included comprehensive control of not only physical training but also nutrition and lifestyle. Unbeknownst to her, the blue "vitamin pills" handed to her after her daily training sessions were actually *Turinabol* (chlorodehydromethyl-testosterone, or CDMT), a highly potent synthetic testosterone derivative. During a time when a typical fourteen-year-old girl anticipates the onset of female puberty, Heidi struggled with a deepening voice, growth of facial hair, and jeers about her appearance on the streets of her hometown. On one occasion as she rode a city bus with her mother, strangers taunted her, suggesting she was a drag queen.

By the early 1980s, Heidi was being groomed by the East German Sports machine, specifically by Stasi (State) Plan 14.25, a government effort through which young athletes were, unbeknownst to them, given high doses of steroid chemicals, with the full complicity of Jenapharm, a pharmaceutical company under the control of the East German government (the German Democratic Republic, or GDR). The scope and irony of Stasi Plan 14.25 was astounding. It is estimated that, at its peak, approximately 1,200 doctors and scientists were employed by the program, and today as many as 10,000 former athletes continue to bear the physical and psychological scars.

In the GDR, the practice of testosterone supplementation followed a pattern of increasing ethical abuses: first it was tried on men, who would presumably experience fewer negative side effects because their "tank" was already full, and then it was tried on women, and then on girls as young as twelve. The greatest danger, of course, is that those with the lowest levels of testosterone to begin with would show the most dramatic effects and would therefore yield more significant data for researchers, and data was what they were after, even at the personal expense of the unwitting test subjects. Heidi's side effects included infertility.

Yet the East German government accomplished its goal; throughout the 1970s and until the fall of the Berlin Wall in 1989, East Germany outperformed even the United States in athletic competition while the world looked on in wonder.

When the Stasi Plan 14.25 was uncovered in 1991, as East Germany fell apart, the trial of its perpetrators took on a Nuremberg-like quality, with many of the defendants claiming to have been "just following orders." Many Germans were shocked that rigorous and systemic state-sanctioned mutilation of its own citizens could occur less than thirty years after the defeat of the Nazis, who were charged with similar crimes and used a similar defense.

Many of those who participated in Stasi Plan 14.25 were able to flee the country before the trial could be completed, and many continued working in sports. Former GDR coaches were not banned from international participation, and they found work in Japan, Europe, and the United States. Unsurprisingly, Turinabol began to appear in the urine tests of athletes from these countries.

By the mid-1970s, before Stasi Plan 14.25 had been exposed, Olympic officials had already caught on to the idea that testosterone and its chemical cousins were being used to dope athletes. In response, the International Olympic Committee (IOC) began testing participants in the games. However, since testosterone is made naturally in all of us, a simple blood test could not determine whether the level showed the athlete's own natural, or endogenous, testosterone level, or whether exogenous supplements had altered the reading. To solve this problem, the IOC developed a test that measured the ratio of testosterone to *epitestosterone* in urine (the T:E ratio). Epitestosterone is one of the breakdown products of testosterone, and while we all have some of it naturally, the ratio can give clues as to whether the testosterone being measured was naturally occurring or administered from outside.

In direct response to the T:E ratio test, the scientists at Jenapharm synthesized epitestosterone and added this to the doses of testosterone they administered to the athletes. They hypothesized that having more "E" in a subject's system would allow them to continue doping with "T" but without affecting the T:E ratio, and therefore allowing their athletes to fool the test. And it worked—so well in fact that the amount of hormones given to the girls was increased higher and higher, far above what was normal, not to mention safe.

In 1986, when Heidi Krieger was twenty years old, she was dosed with an average of just over 7 milligrams of Turinabol per day—well beyond the recommended dose of 1 mg per day for females. The yearly dose added up to 2,590 mg of the drug, a full 1,000 mg more than sprinter Ben Johnson was caught taking, and for which he was banned for life from competing in track and field. Although Heidi went on to win gold medals in the shot put, her appearance—and in fact the appearance of most of the East German female athletes—drew the attention, and the suspicion, of the international sports press. Indeed, these athletes became the butt of jokes worldwide; I can still remember hearing them on TV when I was a young boy staying up late to watch *The Tonight Show* with Johnny Carson.

Meanwhile, as we in the West enjoyed a laugh at her expense, Heidi roiled in confusion as she struggled with her personal life, sexual orientation, and indeed her very identity. In 1997, she completed a female-to-male gender transition and now goes by the name Andreas. Since then Andreas has lived a relatively peaceful life in Magdeburg, Germany, with his wife, Ute Kraus, a former Olympic swimmer and also a victim of Stasi Plan 14.25.

It wasn't until 2000, three years after his gender reassignment was complete and the Stasi records were fully released, that Andreas came to know that he had been a victim of Plan 14.25. He was a key witness in the prosecution of its architects. In 2004, he told the *New York Times*, "They killed Heidi."[41] Ironically, as part of his gender-reassignment treatment, he now requires monthly testosterone injections.

In 2006, Jenapharm paid €9,250 (roughly $12,000) each to 184 athletes (out of several thousand) affected by the plan, thereby insulating themselves from further lawsuits and judgments in the courts. The toughest sentence was handed down to Lothar Kipke, who oversaw the system for the GDR Swim Federation from 1975 to 1985. His sentence amounted to a $6,000 fine and a suspended jail sentence.[42] The case is now considered closed. We'll thus never know the full, lifelong extent of the abuses inflicted upon these young women, who are now well into and beyond middle age.

As we saw in the previous chapter, the variability in how a person responds to testosterone supplementation is driven by factors including fetal testosterone exposure and variations in the sensitivity of the person's androgen receptors. Jenapharm only influenced one-third of what we've been calling the virility triad. Though Turinabol may have been the common fuel in these cases, it was put into many different types of engines. Even as we look at a case in which hormone doping fundamentally changed an individual's identity, the fact that each person has a unique reaction to the same treatment is proof that our destinies are, at least to some extent, forever bound to how we were born.

Part II

THE MANY FACES OF VIRILITY

Chapter Five

LIFTING THE FOG ON LOVE: THE CHEMISTRY OF COUPLING

Our desire for sex is driven by testosterone, but what is the hormone's effect on love? Is staying in love, or even finding new love, possible if testosterone levels drop to zero? Attraction and desire are often a big part of the experience of falling in love—since testosterone drives our desire for sex, can we still find love when it's gone?

One of my long-standing patients, James, has been looking for love for a while. He lives several hours north of San Francisco in a small, bucolic community near the ocean—Mendocino, where in summer the fog can be thick enough to cause the type of seasonal-affective disorder usually seen only in winter. In the fall and spring, though, it's one of the most beautiful places on earth.

For James, however, it had become a lonely place—Janet, his wife of over three decades, died several years back from a neurological condition, when she was fifty-eight and he was sixty. He started coming to see me not long after she died, and I remember him tearing up during our first visit when I asked whether he was married. James had worked in advertising while his wife taught high school English, and like so many retirees from the Bay Area, they'd sold their house in the city and bought a small place up north near the ocean, where they planned to live a long life of love, travel, and quiet living. Her death cheated them both out of that future.

And now James had prostate cancer. Like Aaron, he'd had radiation a few years before, and when that didn't cure him, his levels of PSA (prostate-specific antigen) began to rise. We did hormonal therapy to lower his testosterone levels, and it worked to control the cancer for a couple of years. Then it stopped. The term we use to describe this scenario is one nobody likes—Castration Resistant Prostate Cancer (CRPC)—and we like what it means even less: that the cancer is resistant to the "castrating" effects of androgen-deprivation therapy. But there are lots of ways to treat prostate cancer, and lots of variation in how it progresses.

James's "cancer burden" (the number and size of tumors in the body and how they affect him) was low back then: he had just a few enlarged lymph nodes in his pelvis that extended up into the mid-back area near his kidneys. His bones, which are the most common site of metastatic prostate cancer, were clear of tumors. He asked me to give him a prognosis, and I told him that overall I thought it was pretty good. But cancer is tricky. I usually tell patients that my sense of their prognosis will be better three months after I start a new treatment than before I start it; if the treatment works, the prognosis gets better, and if it doesn't, it gets worse. In his case, I thought James might respond to an old generic drug called *ketoconazole*, which further impairs testosterone production in the adrenal glands and in the cancerous tumor itself. I prescribed it and waited to see what would happen.

My longtime nurse, Jay, gave James instructions on how to take the ketoconazole, when to get his liver monitored (for side effects from the drug), and when to come back and see me. The "keto," as we call it, worked even better than expected in James. Not only did his rising

PSA reverse course, it actually went down to zero, something that happens in maybe 10 percent of patients on ketoconazole. (We can do better than that now with some newer drugs.)

The zero-PSA level held steady, and because James lived so far away (and frankly had better things to do than navigate the windy, foggy, curvy Mendocino coastline to come to my office and hear me say he was doing great), he decided to get his PSA and liver tests done locally every few months and come to see me only once or twice a year. He would call me if his PSA rose or if other symptoms arose.

It went on like this for a couple of years. When I did see him, I felt very grateful to have "another satisfied customer," as we'd sometimes joke.

Still, my conversations with James were always laced with a tinge of melancholy, and one Thursday James said he needed to talk. It was a conversation that he'd put off not only during previous visits, but also until the current appointment was nearly over.

"Dr. Ryan, I want to come off all treatment, OK?"

"Why? You are doing great, and your PSA is zero."

"Well." He paused. "I've been on three dates over the past two years and have met some really nice women."

I settled in my chair—I knew this was going to take a while, so I'd better get comfortable.

"But all three of these women bolted as soon as I told them about my cancer."

Ah, right, the cancer.

"Well, is it the cancer that scares them away, or the fact that you're on hormonal therapy?" I asked. James was still on a drug called *Lupron* to lower his testosterone, and the ketoconazole lowered it even further, the effect being that James's libido was almost nonexistent.

"A bit of both, I think," he replied. "I know these women want a physical relationship, and that's something I just can't give them."

Unfortunately, he was pretty much right. Yes, there's a lot of technology out there that can help low libido and impotence—vacuum pumps, Viagra, and even a drug injected directly into the penis—but it's not the same as the real thing. Plus, James had had both prostate radiation and surgery, and it's safe to say neither of those things was making his sex life any better. While I like to be an optimist whenever possible, I wasn't going to deny reality. I nodded in silent agreement. Given James's history, I concluded there wouldn't be much harm in stopping all treatment, monitoring him closely, and hoping for the best.

Over the next year and a half, James's PSA level was up a bit, but no more than expected. His spirits were good, he grew stronger, and he walked part of the Camino in Spain. He brought me a gift of a twenty-four-ounce bottle of his favorite craft beer. He was happier, if still lonely. Even having stopped his hormonal therapy, I knew there was a good chance his testosterone wouldn't rise much and might never get back to the normal level he desired.

Unfortunately, after that year and a half, James's cancer showed signs of worsening and we restarted his treatment. A while later, the cancer spread to his bones and lungs—luckily it was progressing relatively slowly, and James wasn't in pain. He carried on with strength, humor, and craft beer, but clearly, things were different. He started chemotherapy with an oncologist closer to home.

Four months passed before his name was back on my schedule. I placed my hand on the exam room door and braced myself for a possibly diminished version of my patient, but when I entered I was greeted by an excited James, who literally jumped out of his chair to introduce me to Sharon, his new bride.

Together, the three of us discussed James's cancer journey (Sharon seemed eager to hear it directly from me). I don't think I checked his testosterone that day, but it was probably pretty close to zero, considering the drugs he was taking. Nevertheless, he was acting like a schoolboy in love: laughing and holding Sharon's hand, he had a whole different aura about him.

New love in the face of something as life-threatening as prostate cancer gives us hope that there's more to love than sex, libido, and the hormones that draw us toward potential mates. Most of us believe this, but it's still reassuring to see it happening in front of your eyes. With apologies to Gabriel García Márquez, this is love in the time of cancer. And castration. It must be a strong force, indeed.

So what is the science of love—and how does testosterone fit in?

LOVE *IS* A DRUG

In my year off between undergrad and medical school, I was able to secure a job as a technician in a research lab in the Department of Psychiatry at the University of Michigan, Ann Arbor. We studied the effects of opiate drugs, alcohol, and food intake on certain parts of the brain, as well as some of the chemical and structural components of addiction. One of our tasks was to inject drugs into a rat's brain through a cannula. At that time, a key piece of emerging science was the idea of a common reward or "pleasure path" in the brain. No matter what the stimulus, be it opiate drugs, alcohol, or sex, the same pleasure path would be activated as the final part of the "reward" that made us feel good. We'd inject the rats with opiates to see whether they would eat less fat, or to see what it did to their alcohol consumption. Then we would look at the parts of the brain that were affected and work to translate these results into what we hoped would be interesting and useful observations about eating disorders and addictions, which at the molecular level share some similarities. We were basically studying how a reward through one means (like taking a drug or drinking alcohol) might affect the tendency to self-reward through another (like eating fat).

Today we look at the brain as a type of circuit board. There are different circuits for pain, pleasure, sleep, etc., and different combinations of brain-pattern activation correspond to different emotional, physical, and behavioral states. The "pleasure path" circuit that creates the reward signal in the brain begins in an area called the *ventral tegmental area* (VTA) and drives signaling to an area called the *nucleus acumbens*.

The chemical responsible for all of this pleasure coursing through our brain pathways—no matter what stimulates it—is largely the neurotransmitter *dopamine*. Drugs, alcohol, and sexual attraction all increase dopamine levels, so yes, it turns out that new love *is* sort of like drug addiction. Specifically, at the level of the VTA and nucleus acumbens, the two conditions look pretty much the same. Testosterone and dopamine are tightly linked, and in fact have a bidirectional relationship, which means that each influences the other. Testosterone increases dopamine activity in both men and women, and it's likely that much of the pleasure that people on testosterone supplements experience is due to an increase in the dopamine coursing through the brain's pleasure pathway.

Dopamine does not have the same effect in both men and women, however. In women it leads to testosterone production, whereas in men testosterone levels go down—which may explain why we think of men becoming "docile" when they're falling in love. A man in love may be less likely to abuse drugs and alcohol and less likely to chase after new romantic conquests, as his brain doesn't need another reward stimulus. Love is his natural heroin. It is its own special brand of neurochemical wonderfulness.

So why did evolution program us so that women who fall in love get a boost in testosterone while men experience a drop? The answer is pretty simple. To put it in terms of evolutionary biology: for the man the hunt is over. When he's in love, his thoughts turn to nurturing and affection, not competition and domination. For the woman, on the other hand, this newfound affection needs to be translated into something else, something that will help preserve the species—namely, reproduction. The bump in libido produced by the rise in testosterone inspires her to accomplish that goal.

It may come as a surprise, but it's true that although testosterone affects the desire for sex and the urge to reproduce, my observation is that it has relatively little to do with love. When you remember that testosterone is the chemical of aggression, dominance, and reduced empathy, it makes perfect sense. It is rare indeed for my patients who undergo testosterone-lowering therapy to lose affection for their romantic partners. In fact, I've had many patients, like James, who have fallen in love while taking these treatments. Testosterone governs many simple actions and emotions, but it can't be responsible for something as complex as falling in love.*

Here is something interesting: before we initiate a dramatic reduction in testosterone to treat prostate cancer, the number-one worry among my patients is loss of libido. But while it's true that the libido is doused after we start treatment, the reality is that many patients don't end up caring about it too much; the trend suggests that testosterone affects not only the libido but perhaps also *caring about the libido*. This scenario is in stark contrast to a patient with a robust libido struggling with erectile dysfunction—in that case, the desire is there but the mechanics aren't—and here again we see that testosterone affects the *desire* for sex but not the ability to get an erection and actually *have* sex.

Considering how much mental space is taken up by thinking about sex, the idea of caring less about libido may actually open doors for some men, allowing them to concentrate instead on any number of other things, including developing a deeper romantic relationship that

---

*Love is so complex, we often think of it as a uniquely human phenomenon, but we're not entirely sure that's true. Although my sensitive, romantic side wants to say that it probably is uniquely human, many animals make dopamine and have this same pleasure circuit—including those very unromantic rats I studied in the lab at the University of Michigan.

exists independent of sex. I've heard this notion expressed by some of my patients over the years, and in fact it's an old idea, and one that has been written about before. Consider this scene from Plato's *Republic*, written more than two thousand years ago:

> I [Cephalus] was once with Sophocles the poet when someone asked him, "How do you feel about sex, Sophocles? Are you still capable of having sex with a woman?"
>
> He replied, "Be quiet, man! To my great delight, I have broken free of that, like a slave who has got away from a rabid and savage master."
>
> I thought at the time that this was a good response, and I haven't changed my mind. I mean, there's no doubt that in old age you get a great deal of peace and freedom from things like sex. When the desires lose their intensity and ease up, then what happens is absolutely as Sophocles described—freedom from a great many demented masters.[43]

COUPLING FOR THE LONG HAUL

New love, and the feeling of intense pleasure it brings us, is fleeting. What of love that lasts decades? My wife and I started dating in college, and although we were only a year into the relationship by graduation, we were committed enough to each other that we stayed together when I went off to work in the lab in Ann Arbor and she was accepted to join the Jesuit Volunteer Corps in California. While I was spending quality time with rats—cannulating their brains, weighing their feces, and counting how many calories they got from fat versus carbohydrates—she was working at a community-service organization for little to no money, but reaping spiritual rewards that still inform her dedication to social justice today. We both benefited from the work we did that year, and we were lucky enough that I was able to visit her every eight weeks or so, thanks to a special "recent graduate" deal through Continental Airlines that allowed me to fly round-trip for something like $160. We also talked on the phone every night after work.

Despite the distance, our relationship did not suffer; we had formed a solid bond and we made an effort to stay connected. Interestingly, there is some science behind this phenomenon. Single men have been shown to have generally higher testosterone levels than those in relationships—probably because, as we discussed, they are still in "compete to reproduce" mode—but contrary to what you might guess, the testosterone levels of men who are in committed long-distance relationships are no different from the levels of those in committed same-city relationships.[44] This may seem surprising, as it feels counter to the idea that testosterone levels would be affected by the proximity of a partner—perhaps driven by *pheromones*,* which must be smelled to work their magic. My wife and I have now been married almost twenty-five years and we both look back very fondly on that year apart, considering it something of a test we passed with flying colors. I am reminded of that year every time I look at my wedding ring—on the same finger as the V-shaped scar from a rat bite. (FYI, rats don't like a fat-restricted diet and can get a little ornery.)

If I were in charge of prenuptial counseling sessions, I would require each engaged couple to spend a weekend with a pair who has been married for fifty years or more. Imagine what they could learn from these long-term survivors of (and thrivers in) marriage! I'm certain they'd learn that marriage isn't always easy, but I bet they would also see a lot of affection. What they would probably *not* see is much of the behavior we classically attribute to testosterone—aggression, domination, and lack of empathy and generosity. Testosterone may be the

---

*Human pheromones are chemicals given off by the body for the purpose of attracting a mate.

hormone that attracts mates, but it doesn't do us many favors in keeping them: psychologists have determined that the failure of relationships is tied to higher testosterone levels.[45] Research done in the late 1990s* explored the connection between heterosexual marriage quality and the testosterone levels of the man. In short, the studies suggest that men with higher testosterone levels are more likely to get divorced, have extramarital affairs, and be violent with their spouses. (Men with higher testosterone levels are also more likely to work in manual labor, fight, get injured, and die young.)[46, 47]

As for the women in male-female partnerships, one of the more interesting findings of recent research is that the testosterone level of the *female partner* may be a significant determinant of the overall satisfaction of a couple as well. The most remarkable detail is that there's a negative correlation between the woman's testosterone level and the self-reported satisfaction of both members of the couple. It turns out that men, too, were more satisfied with female partners who had lower testosterone. In fact, in one recent major study, the woman's testosterone level had a greater impact on marriage satisfaction than did the man's.[48] So, while the libido of a woman may be driven by testosterone, and therefore one might assume that a woman with high testosterone = high libido = high sexual satisfaction for the male partner = greater overall satisfaction for the male partner, the data shows that simply isn't so. Nor does it appear that the two balance each other out—in other words, a high-testosterone man isn't happiest with a low-testosterone

---

*So much of this interesting work was done in the 1980s and 1990s, when researchers didn't have access to what we now know about genetics, such as the variations in the androgen receptor that may affect what testosterone does in the body, not to mention some of the more sophisticated brain-imaging techniques we use today to get a peek inside these complex processes. Nevertheless, even when reducing the analysis down to just one variable—testosterone levels—we can learn much about how these chemicals work.

woman, and vice versa. It appears that the greatest happiness is when both are at the lower end of the range.

The cause and effect of all of this has not yet been worked out. Are these marriages bad because the testosterone is high, or is the testosterone high because the marriage is bad and each partner's "attraction" instinct is kicking in as an effort to save the relationship? Either way, a consistent and pervasive theme we have seen is that testosterone reduces nurturing feelings and actions. In the long haul of a fifty-year or even a five-year relationship, what wins out is the nurturing, mutually respectful, and stable qualities that are the hallmark of lower testosterone.

Recently some provocative data has been published on how heterosexual relationships are affected by *androgenicity*, which is the presence of physical characteristics that are typically male, such as abundant body hair, male sex organs, and larger muscle mass. High androgenicity is linked to both higher serum levels of testosterone and more-active androgen receptors, based on varied lengths of the *CAG repeat*.* The data showed that divorce rates were higher among men with both low androgenicity *and* high androgenicity, while the more stable relationships were in the middle.[49] Here we have another bell curve.

But why did both the low and high end have similar results? The answer may be that the behaviors of people with high versus low androgenicity/testosterone are qualitatively different, meaning that while the outcome is the same, the reasons behind the outcome are not. Low testosterone has been linked to depression (more on this in a later chapter), and perhaps the relationship problems of those on the low end of the androgenicity spectrum are related to factors like this, while at the higher end there may be more abuse and higher rates of infidelity.

---

*CAG stands for cytosine, adenosine, and guanine, which are nucleic acids found in deoxyribonucleic acid, or DNA. The term "CAG repeat" describes a sequence of these DNA segments at the beginning of the androgen receptor gene that governs how active the androgen receptor will be. The length of CAG repeats varies by individual.

BENDING AND BONDING

According to the folks at Guinness, the world record for the longest marriage was set by Herbert and Zelmyra Fisher of James City, North Carolina. On February 27, 2011, the day Mr. Fisher passed away at age 104, they had been married 86 years and 290 days. Their advice for other couples? "Remember, marriage is not a contest; never keep a score. When disagreeing, learn to bend—not break!"

Couples who have been together for many, many years make for sweet stories, but is a long marriage in our biology? It's possible my patient James's hormonal state makes him especially well equipped for long-term affection. Putting the cancers aside, the reality is that the low-testosterone milieu he has been living in may affect his testosterone-to-estrogen level in such a way that his brain is now hormonally similar to that of an older female brain. This situation may enable the emergence of a greater effect from oxytocin. Louann Brizendine, in her book *The Male Brain*,[50] offers this as an explanation of the affection and generativity that may be seen more in older men than in their younger counterparts. An altered balance of oxytocin to testosterone may also underlie some of what Aaron observed of more affection and greater empathy after his hormonal therapy.

The term "successful marriage" is used frequently, but how is it defined? If you think it means simply not getting divorced, consider how many marriages have a good run before they end, sometimes amicably, and how many others stand the test of time but aren't what anyone would call happy. A concept attributed to the cultural anthropologist Margaret Mead goes something like, "Everyone should have three marriages: one for sex, one for kids, and one for companionship. All three can be with the same person." While I suspect most people still at least try to find one person to fulfill all their needs, those needs, and the priorities one gives to them, are not only different from person to person, but they also might well change over time. I have no idea when the phrase "'til death do us part" came into vogue for wedding ceremonies, but I bet death parted couples a lot earlier back then. With advances in medicine and nutrition, the average life expectancy is longer now, and it makes sense that people change over the decades when, hundreds of years ago, they might already be dead.

Just as relationships change over time, the biology of connection, affection, and bonding do as well. If anything, one would expect that as a marriage matured the hormonal fluctuations would quell and the emotional bonds would become stronger, but even if the first part of this is true, the second is obviously not always the case. "Silver divorces" are quite common these days. Many rightly feel empowered, not enfeebled, as they enter older age. Maybe when the hormonal fog lifts, we can see more clearly. Or maybe "silver divorces," long-distance relationships, and men who find new love against the hormonal odds are all examples of the ways we can override or even influence our own biology.

James and Sharon won't break the record held by the Fishers, and because of the cancer and the effects of his hormonal therapy, their marriage may be a little different from most. Regardless, James found what he was looking for—someone to be at his side in the fog. For the good times and the bad; the craft beer and the chemotherapy.

Chapter Six

THE UNQUIET SUNSET: TESTOSTERONE AND ALZHEIMER'S

I met Warren in my first year of medical practice. His daughter Kimberly was a friend of mine from school and also a physician, and when I ran into her at a conference she asked if I might be able to help with her dad. He was seventy-six years old, and had been diagnosed with and treated for prostate cancer a decade earlier. He had been treated with hormonal therapy, but after the shots were discontinued his testosterone had never recovered to a normal level and, according to Kimberly, he was miserable as a result. The biggest source of her frustration was that though Warren was now cancer-free, none of his doctors would supplement his low testosterone for fear that the cancer would recur.

He came to see me at my clinic. The man I met not only didn't appear miserable, but was congenial and smartly dressed—creased and ironed khakis, short-sleeved white Oxford shirt, loafers, and a thin tie with a tie clip, the ensemble topped off with retro glasses worn non-ironically, both a pen and a spare in his chest pocket, and a salt-and-pepper buzz cut. Think 1960s NASA, like Ed Harris in *Apollo 13*. It looked like he'd put in a fair amount of effort, and I not-so-secretly love when patients wear ties to come see me. It's an old-school sign of respect, the complement to the fact that without a tie I myself feel underdressed when seeing patients in the clinic—this in spite of the fact that we now know that a doctor's tie is basically nothing more than one long bacterial culture medium.*

Warren was a retired internist, and we hit it off instantly, as is usually the case with patients who also happen to be physicians and are from the same generation as my dad. We started the appointment with the usual small talk, and then I moved on to direct questions about fatigue, depression, stamina, and libido. As we continued, it seemed that he was a bit depressed. Mostly it was low energy, decreased stamina, and an understandable frustration over those things. There were no thoughts of suicide, self-harm, or other hallmarks indicating a more severe problem, and I determined it was a mild enough case that there was no need for him to see a psychiatrist or go on an antidepressant.

It seemed likely that low testosterone was indeed the problem, and I told him as much. I had him schedule an appointment for labwork the next week so I could check his total testosterone level, and sure enough, it was very low, in the "castrate" range at 34 nanograms per deciliter. The normal level for a man his age would be 250 ng/dL or higher. Warren's tests showed that he was still producing a small amount of testosterone in his adrenal glands, which aren't affected by

---

*Studies have shown that about 30 percent of doctor neckties were shown to harbor potentially dangerous bacteria. In 2006, the United Kingdom's National Health Service recommended that UK physicians stop wearing neckties to halt the spread of MRSA, an antibiotic-resistant staphylococcus. Such practice has not yet been endorsed here in the United States.

the medications he had taken, but it clearly wasn't enough to have him feeling his best.*

The likely cause of Warren's low testosterone was that his testicles had simply shut down as a result of the androgen-deprivation treatment he'd undergone a decade earlier; they'd atrophied and couldn't recover their ability to produce testosterone, even after the shots were stopped. When Warren was first treated, in the mid-1990s, there was less understanding of the long-term effects of this treatment and greater worry that stopping treatment would allow the cancer to recur, and therefore many men received longer courses of treatment than they would today.† In Warren's case the treatment had lasted only a couple of years, but there he was, ten years later, still feeling the effects. It might seem like a fine trade-off for not having cancer, but we can't dismiss a decade of inadvertent castration and the toll it took on his quality of life.

Depression is common in older men and can be a particularly obstinate problem. Clinical research studies demonstrate with some consistency that men with low testosterone levels who suffer from depression will benefit from testosterone supplementation, and studies that compared testosterone replacement to a placebo also showed impressive results.[51] The pre-depression symptoms (sluggishness, listlessness, etc.) that many men face as their testosterone goes down are a major driver of the commercial success of "low T" clinics and their

---

*In the 1940s, when hormonal therapy for prostate cancer was in its infancy, doctors experimented with surgical removal of the adrenal glands. Unfortunately, the results were declared a failure, as patients quickly died because their bodies could no longer produce cortisol, which is required for the maintenance of several critical systems, including heart function and blood sugar levels. These same risks are present when a person stops taking prednisone (basically cortisol) or other steroid medications; quitting cold turkey can trigger secondary adrenal insufficiency, so the drug must be tapered slowly in order to allow the adrenal glands to "wake up" again.

†Nowadays many men, like Aaron, are treated by alternating one year off and one year on.

ubiquitous radio and TV commercials, and many doctors agree that men who are depressed should at least have their testosterone levels checked to see whether this may be a problem with an easy solution. Quick fixes are rare, but they are always gratifying to both patients and doctors.

As I reviewed Warren's history, I saw little potential harm in giving his levels a boost. It was out-of-the-box thinking at the time, but today is more widely accepted. He'd had a PSA of zero and low testosterone for ten years, and considering the grade and stage of his earlier prostate cancer, it was unlikely that he'd experience a recurrence, though of course we would still monitor him, just to be safe. I told him all of this when he returned the next week. Then I prescribed topical testosterone gel, and off he went.

He returned a month later, doing well. There hadn't been a dramatic transformation, but he certainly felt better—more spry and energetic for sure. And so it went, month after month. Every time I saw him there was a bit more improvement. His spirits were good, and he confessed that his mood had improved enough that he had been enjoying, as he called it, "watching the ladies." Comments like that are usually a sign testosterone levels are approaching normal, and his test results confirmed it.

We monitored Warren's testosterone level as it climbed up from the low point of 34 ng/dL into the 150s, then 200s, and up to about 385 ng/dL. This put him in the normal range for a younger man, but not too high for a man who by then was approaching eighty. Life was good. Most important, there was no sign that we had stimulated the cancer to grow.

Warren enjoyed the positive effects of testosterone supplementation for two years, but then things took a turn. One day I received a call from a colleague in neurology, who told me Warren had been referred to them with some worrisome symptoms: memory loss, misplacing things, forgetting certain people and places, etc. The worry, of course, was that he was developing Alzheimer's dementia. And, unfortunately, he was. I'd been called because, on top of all his other symptoms, Warren had become sexually inappropriate in public. It started with rude comments and physical gestures to women, and his loved ones were shocked and embarrassed. This was completely out of character.

"Could we discontinue the testosterone?" the neurologist asked coldly, as if it were a decision for which I would need to muster all my medical training. Of course we could, and did, and in a month or so Warren's sexually explicit behavior had mostly stopped. But I still felt more than slightly responsible for Warren's misbehavior, and I even wondered whether the testosterone had somehow accelerated the Alzheimer's disease.

Once Warren stopped taking testosterone, I saw him one more time, briefly, but our therapeutic relationship was basically over. Sadly, discontinuing the testosterone had no effect on the pace of his Alzheimer's disease, and Warren's dementia continued unabated. Soon he was in a nursing home, and within two and a half years, he was dead.

TESTOSTERONE ON THE BRAIN

Two decades later, we have epidemiologic and clinical data that helps clarify the role of testosterone in cases like Warren's. As it turns out, if anything contributed to Warren's development of Alzheimer's, it was the ten years of *low* testosterone, not the replacement gel that brought his levels back up. My prescription may have fueled his libido in a way that contributed to his sexually inappropriate behavior, but this behavior is, unfortunately, common in Alzheimer's patients.

As we control diseases like cancer, more people are living long enough for dementia to take hold; today, more than 5 million Americans live with Alzheimer's disease. Dementia affects women at a ratio of six to one over men, and this isn't just because women live longer; they have a higher *incidence* of the disease as well as a higher *prevalence*.* Although the difference in Alzheimer's incidence between men and women does not prove that testosterone protects the brain from Alzheimer's (i.e., that it is because women have lower testosterone that they are more at risk), the data is at least consistent with that hypothesis. There is also some fairly compelling data showing that the type of androgen-deprivation therapy many prostate cancer patients undergo may increase their Alzheimer's risk.

There's little doubt that testosterone helps prop up the brain of the aging man, and given the potential link between Alzheimer's dementia and the treatments I prescribe—and in light of what happened to Warren—I began delving into the research on the relationship of testosterone to Alzheimer's disease. I was blown away by what I found. For all that I already knew about the increase of empathy and "niceness" in men with lowered testosterone, there were also valid concerns about whether low or declining testosterone could contribute to the impairment of executive functions such as memory. So what *are* the effects of loss of testosterone on the human brain? Although it sounds worse than it actually is, one of the more shocking effects is that the brain actually shrinks with testosterone depletion. This is due to a decline in the *volume* of gray matter. Think of it as neurons getting "thinner," not necessarily being damaged or dying. In some areas of the brain the gray matter volume decreases by over 50 percent.[52] The exact cause and effect of this phenomenon is not known. It should also be mentioned that brain shrinkage is very common in aging—so much so that radiologists often mention in passing "age-associated volume loss" when they report on the brain MRI findings of a patient. Nevertheless, the data suggests that loss of testosterone may accelerate this phenomenon.

---

*Incidence is the rate at which a disease develops (e.g., new cases per year), whereas prevalence is the total number of people living with a disease, regardless of when they were diagnosed. These two terms are frequently confused.

Up to now, when I've used the term "testosterone levels," I've usually been referring to the concentration of testosterone in the blood;
that's not the whole story, however, as the hormone also concentrates
in certain organs that depend on it to function. The prostate is a classic example of a testosterone "sink," where concentrations are much
higher than in other nearby organs, like the kidneys. The brain is also
a testosterone sink, and in addition to trapping and preserving some of
the testosterone circulating in the blood, also makes some of its own.*
The testosterone made in the brain stays in the brain and is active only
there. We can see this happening in the brain when we look for not just
levels of the hormone itself but also mRNA† of the genes that code
for the enzymes responsible for producing testosterone. The key point
here is that the presence of this mRNA in brain tissue suggests that the
hormone is being made there—as opposed to being made solely in the
testicles and reaching the brain through the circulation.

Testosterone levels in the brain, rather than in the blood, can be
measured by tests performed directly on brain tissue (thus making it
hard to do with a living subject). Researchers in California took post-
mortem measurements of testosterone levels in normal brains of
people who had died with neither prostate cancer nor Alzheimer's disease, and they found, perhaps not surprisingly, that testosterone levels
in the brain were lower in older men than in their younger counterparts. Moreover, the difference wasn't subtle. By age ninety, men lucky
enough to still be around will have about *one-eighth* of the brain testosterone they had at fifty.[53]

---

*Similarly, when prostate cancer becomes resistant to testosterone suppression
and grows, it is often doing so by making its own testosterone. It's an adaptation
by the cancer that allows it to survive in an otherwise low-testosterone environment. That's what cancer does—it finds a way to survive, by hook or by crook.
†Messenger RNA is a component of RNA, which carries a segment of DNA to
various parts of the cell for processing.

Hormones protect the neurons and the brain in general, so the loss of the hormones, whether through treatment for prostate cancer or the normal aging process, may contribute to the development of Alzheimer's disease. In men, loss of testosterone is a potential culprit.* And yet, despite the sudden and dramatic loss of testosterone in men who undergo hormonal therapy for prostate cancer, only a minority go on to develop Alzheimer's disease. One study that followed a group of men for approximately ten years found that about 4 percent of subjects in their seventies who were *not* on hormonal therapy were diagnosed with Alzheimer's, compared with the 6 and 7 percent of those who, like Warren, underwent androgen-deprivation therapy.[54]

Sometimes when this type of study gets reported in the media you see headlines like "Alzheimer's Risk Increases by Fifty Percent with Hormonal Therapy." This is technically true—an increase from 4 percent to more than 7 percent *is* over 50 percent—but it should not be confused with an increase of 50 *percentage points*, which would be a jump from 4 percent to 54 percent. The second set of numbers shows an *absolute* 50 percent increase, while the first set of numbers—the ones representing the reality of the situation—show a *relative* 50 percent increase.

Many of the health and science reporters I've talked to over the years work hard to clearly express scientific subtleties like the one above, but it's not easy; we all too often see journalism that seems to have been the work of sloppy reporting, a lack of understanding, or just blatant sensationalism. For example, it was particularly disappointing to read a *Wall Street Journal* article about the link between androgen-deprivation therapy and Alzheimer's risk that said, "Of those . . . treated with anti-androgen therapy, . . . they had an 88 percent higher risk of being diagnosed with Alzheimer's disease in the next three years than those

---

*Interestingly, testosterone levels in the brains of women appear to be stable throughout the aging process, as are levels of estrogen.

who weren't." The numbers simply weren't true, and in addition, this particular article conflated two controversies—the link between hormones and Alzheimer's, and the overtreatment of early-stage prostate cancer—suggesting that doctors were using the treatment in patients who didn't need it and thus *causing* Alzheimer's. Perhaps worst of all, the single graph illustrating the article showed the worldwide growth of money spent on androgen-blocking drugs, implying there was some sort of financial incentive behind how doctors were choosing to treat their patients.[55]

Patients sometimes bring clippings and printouts of such articles to their visits, and in some cases demand their doctors discontinue treatment on the basis of what they've seen in popular media. Even if they agree to continue treatment, it can take a long time to change a patient's mind on topics printed in trusted sources such as major national newspapers. Therefore, it's important that, as consumers of health-care media, we all look at and try our best to understand the data, rather than just relying on attention-grabbing headlines.

With that said, although this link between loss of testosterone and increased risk of Alzheimer's may be relatively rare, it is also likely real.

AN ALZHEIMER'S PRIMER

In the mental-health and neuroscience world, the several types of dementias are now called "major neurocognitive disorders," and Alzheimer's is by far the most common, accounting for three-quarters of all dementias seen in older adults. A patient is diagnosed with Alzheimer's when he or she undergoes a significant decline in one or several cognitive domains, including learning, memory, executive function, social cognition, and language. These deficits must progress slowly over time—they don't occur overnight—and they can't be explained by some other diagnosis, such as an injury, a brain tumor, schizophrenia, or a metabolic disorder. And, to be clear, when a prostate cancer patient is diagnosed with Alzheimer's, it is *not* because the cancer has affected his brain; the brain disorder is linked to the treatment, not the disease.

Now that we have an idea of the behavior of a person with Alzheimer's, let's take a look at what is happening on a molecular level. What changes occur in the brain? And what does testosterone have to do with it?

The main feature of Alzheimer's disease is the buildup of protein fragments within the brain. Two types of *plaques* are visible under the microscope: *beta-Amyloid plaque*, a deposit seen outside the brain cells but within the matrix that surrounds it, and *neurofibrillary tangles*, which are found within the brain cells. The mystery of the disease is how and why these protein deposits are formed, and the key to finding a cure is learning what can prevent their deposition and/or what, if anything, can be done to treat or remove plaques once they are there. Nobel Prizes will be waiting for the team of scientists who can definitively answer these questions.

The current thinking is that the plaques are essentially aggregations of cellular-protein "garbage" that accumulate over years as the result of normal processes, including cell metabolism, cell injury, or some other type of cellular stress. What we do know is that somehow androgens, and to a lesser extent, estrogens, help clear these garbage protein deposits—that is, until they don't.

If you place neurons in a petri dish filled with, among other nutrients, testosterone, they will survive and appear healthy. If you then try to grow the same neurons in a petri dish *without* testosterone in the nutrient mix, you will see plaques form of protein clumps called Amyloid beta, or Aβ. If this same scenario proceeds unabated in the brain, the patient may develop Alzheimer's disease. Recent research suggests that this process may be due to the failure of an enzyme called *neprilysin,* which degrades Aβ. The gene for neprilysin contains an androgen-receptor response element.[56] This is critically important because it means that the androgen receptor itself turns on neprilysin, which allows it to degrade Aβ. And what turns on the AR? Testosterone, of course. So that may be the link: less testosterone leads to less neprilysin, and less neprilysin leads to less Aβ degradation, which then allows the buildup of plaque in the brain and brings on the symptoms of Alzheimer's.

Another link that is receiving a lot of attention is the connection between Alzheimer's and *apolipoprotein E* (ApoE), a protein that carries lipids (fatty acids and their derivatives) in the blood. These lipids may be critical for the day-to-day repair of neurons in the brain, and genetic variation related to ApoE among individuals may explain why some people are at risk for Alzheimer's and others are not. Humans have one or a mixture of ApoE types, and research suggests that one specific type, ApoE4, may be less efficient in providing lipids for brain repair. Humans who carry ApoE4 (either one or two copies on their genes) appear to have a higher risk of developing brain problems arising from a variety of causes, be they trauma, reduced blood flow, or infection with HIV. In short, the brain is less protected against these things and therefore at greater risk of declining because of them.

ApoE and testosterone are also linked. For reasons that aren't clear, the androgen receptors within the cells of ApoE4 carriers generate a weaker bond with testosterone and related hormones than in ApoE4-negative individuals. Mice who have even one copy of ApoE4 have been shown to suffer cognitive problems when treated with *flutamide,* a prostate-cancer drug that blocks androgen receptors. Mice with ApoE3 or E2 do not.[57]

Our research team at UCSF and through the Prostate Cancer Foundation is now testing whether men with ApoE4, among other genes, are more likely to experience cognitive problems when we reduce their testosterone to treat prostate cancer. If our hypothesis is correct and there *is* a link, we may move one more step closer to translating epidemiologic findings into patient care, and that will hopefully lead to treatment or prevention strategies down the road.

THE FUTURE OF ALZHEIMER'S RESEARCH

In light of what I know now, I think back to the many possible ways Warren's treatment might have proceeded. Although we've seen how the presence of testosterone might help *prevent* Alzheimer's, we don't have evidence showing it is effective as a treatment for the disorder once it's present. And yet, should I have kept the testosterone treatment going *after* he was diagnosed with dementia, in order to slow down the process? Although one promising study by a research group in Seattle showed that men with either Alzheimer's or a condition known as *mild cognitive impairment* (MCI), which generally precedes full-blown Alzheimer's, benefited from short-term testosterone supplementation[58] (specifically, it increased their spatial cognition), we have other evidence that suggests long-term treatment is likely to override those benefits because it drives behaviors like impulsivity and sexual aggression. It has been known for some time that men who already have Alzheimer's disease and who also have higher testosterone levels are more likely to act aggressively,[59] as Warren did, and a more recent study showed that women with Alzheimer's who behaved aggressively were those who had higher testosterone levels as well.[58, 60] It's pretty hard to argue that a *possible* benefit to some specific cognitive skills is worth that risk.

Pathologists and researchers who study Alzheimer's disease describe its severity according to a classification method called *Braak staging*, named after the German anatomist Heiko Braak, who was the first to show that increased concentrations of neurofibrillary tangles were associated with more severe cases of Alzheimer's disease. The concentration of neurofibrillary tangles can be classified as Braak stage I through VI. The more tangles, the higher the Braak stage. In 2009, a group of researchers in Italy performed autopsies on both men and women who had died with Alzheimer's disease, and what they found was striking to me as a researcher of prostate cancer. The autopsies showed that in the individuals with higher Braak stages, the molecular machinery for the production of testosterone had actually *increased*.[61] It was as if, in response to its degeneration, the brain *tried to make more testosterone*. This molecular mechanism is very similar to the mechanism by which cancer may begin to produce its own testosterone in order to fuel its own growth. In this case, the brain seems to be trying to protect itself. This research is still in the early stages, and other studies are ongoing, but evidence in favor of the link between neurofibrillary tangles, testosterone, and Alzheimer's is growing, and it is exciting to think that this might help us learn a bit more about surviving both Alzheimer's and prostate cancer.

Although testosterone may protect the brain, some characteristics of androgen receptors seem to work in another direction. The brain is filled with ARs, and they are particularly abundant in the hippocampus and other parts of the brain that control cognition and memory. As we have seen, ARs also govern most of the biological processes that we associate with testosterone. Researchers working in Nottingham, England, found that faster (and therefore more receptive) ARs (the result of having shorter CAG repeats) were more common in men with early-onset Alzheimer's disease (although there was no association in women with the disease).[62] Later, they showed that the greatest risk for Alzheimer's was in men with a combination of short CAG repeats and low serum-testosterone levels.[63] Given that low testosterone seems to be implicated in Alzheimer's, one might have expected to see the greatest risk in those with low serum testosterone and *long* CAG repeats, meaning the ARs were slower and therefore less receptive to what testosterone was present in the system. The fact that this is not the case highlights the complexity of the process and tells us there must be at least one piece of the puzzle we haven't yet figured out. My theory is that the brains of the subjects with short CAG repeats were more dependent on testosterone levels than those with long CAG repeats and thus more susceptible to the effects of its decline, but the research has not been definitive.

So where does that leave us in our treatment of prostate cancer with androgen-deprivation therapy? On one hand, the data suggests that testosterone might help prevent Alzheimer's, but on the other hand, it hasn't been shown to reverse the course of the disease, and there are a host of negative effects that occur with too much testosterone in patients who already have dementia. For now, exercise, nutrition, and keeping the brain active are the best we can offer as a recommendation to prevent Alzheimer's and mild cognitive impairment. If we can identify risk factors, as we hope to do in the course of our research on ApoE4, we might then alter the course of care for carriers of that gene, for instance with shorter durations of hormonal treatment, early referrals to neurology, and/or the targeted inclusion of brain-stimulation exercises.

In the meantime, we're stuck in an ethical quandary of sorts: With no cure and not many treatment options, should doctors inform individuals that they are at higher risk for Alzheimer's disease? Should we even test for risk factors in the first place? I can imagine a future in which I have this potential conversation with patients: "If you undergo hormonal therapy, you might get Alzheimer's disease; if you don't, you might die of prostate cancer." Understanding which men are at highest risk of these complications will help us navigate decisions about treatment. This is the work we're doing now, and the fact that our research on prostate cancer might cast a ray of light on the mechanism of Alzheimer's is not only gratifying, but also a powerful indicator of just how vast and complicated the influence of our hormones can be.

Chapter Seven

THE ALOPECIA PARADOX: OF HAIR AND HORMONES

Most people have a sense that genetics governs hair patterns—the evidence shows up in every family photo—and we are also aware that baldness commonly occurs with aging. About 40 percent of men have some noticeable hair loss by age thirty-five, and that number rises to 60 percent by age sixty and 70 percent by age eighty. Many of us also understand, albeit vaguely, that testosterone seems to play a role. While laypeople are likely to be familiar with the phrase "male pattern baldness," the clinical term is actually *androgenic alopecia* (AGA), a name that highlights the role of androgen hormones, including testosterone. Hair loss and virility stick together—literally. Not only because it is intimately related to testosterone, but because, as we'll see later, the gene for baldness, such as it is, is linked to the androgen receptor. You might even say that baldness, by advertising a man's high levels of testosterone activity, is a mark of virility. So why don't we think of it as a desirable trait? Is there a difference between "bald" and "balding"? How far must a hairline recede before it crosses the line into baldness? Is a man with only one hair on his head bald?

We are in the territory of Greek philosophy here. It's called the *sorites paradox* and it's attributed to the logician Eubulides of Miletus. (No disrespect to the ancient Greek philosophy community, but this name sounds like a gallbladder disease.) There are a variety of examples of *sorites paradoxes*—sometimes referred to as the "little by little" paradox—and among the most famous is the "heap of sand" example. (The word "sorites" derives from the Greek word for "heap.") Clearly, the philosopher suggests, one grain of sand is not a "heap." Nor are two, or three, and so on. However, at some point it does become a heap. While we can't say definitively when that level is achieved, we do know it when we see it. In our case, the sorites paradox describes the fact that it's hard to say at what point a man can be considered bald. For the purposes of this chapter, we'll use the word "bald" to indicate someone with significant hair loss.*

Now, let's take a look at what testosterone actually has to do with baldness—and the paradox in our perception of the relationship between hair and virility.

HAIR PARADOX

Perhaps the most perplexing connection between testosterone and hair growth in men is that the hormone stimulates *beard* growth, even as it leads to baldness on the top of the head. Is this yet another paradox? The answer reveals itself in a basic understanding of how hormones work once they enter cells and bind to their receptors. Recall that the business end of the testosterone system is the interaction between the

---

*Applying Greek logic to the analysis of hair is the perilous result of a liberal arts education run amok. Thank you, Jesuits!

hormone and the androgen receptor—a key-in-lock relationship. These receptors, meanwhile, are found in different kinds of cells (for example, a face-skin cell or a scalp-skin cell). In other words, it's not only about the key (testosterone) and the lock (the androgen receptor) but also differences in the rooms behind the doors—what is set free when the door is successfully unlocked.

Androgen receptors are plentiful in the skin of both the scalp and the beard area in most men, but the effects of their activation by testosterone differ drastically. The difference has nothing to do with the level of testosterone or the reactivity of the receptors; it is a function of how the receptors in each kind of tissue are programmed to respond. In face-skin cells, androgen receptors stimulated by testosterone respond by initiating a cascade of growth, promoting production of a chemical called *insulin growth factor* (IGF); you might say the testosterone "key" unlocks the door and lets out IGF. Conversely, in the scalp, testosterone activates processes that repress hair growth, by "letting out" a substance called *transforming growth factor*, reducing the diameter of follicles to cause thinning, as well as promoting the production of oil (sebum) that can clog hair follicles and stop hair from growing.*

To complicate matters even further, some of these processes may even be sensitive to a person's thoughts and emotions. In a now classic single-subject experiment (and a fun read!) published in 1970 as an anonymous letter to *Nature*, the world's leading scientific journal, a man demonstrated how the *anticipation* of sexual activity alone can stimulate beard growth.[64] His data showed that even the presence of a woman, in the absence of sexual activity, could exert a similar effect. His letter began, "During the past two years I have had to spend periods of several weeks on a remote island in comparative isolation," and

---

*These same androgen-induced oil excesses also lead to typical acne during the teenage years, when testosterone is known to surge. One of the drugs commonly used to treat prostate cancer, Bicalutamide, is an androgen-receptor blocker that was originally developed to treat acne.

one can only imagine why that might have been; perhaps he was a lighthouse keeper or a guard on an island prison, if such things still exist. Whatever the reason for his presence there, this gentleman certainly made the most of his downtime on the island. Many scientists work their entire careers and never publish in *Nature*. I still don't know why this was published anonymously!

At any rate, while on the island, our anonymous shaver meticulously measured the weight of his beard shavings on a daily basis. During his working weeks on the island, he would return home each Friday to civilization—and to a sexual relationship. What he documented was that as the weekends approached, the weight of his daily shavings would increase. In periods when he was not working on the island and instead was sexually active on a more regular ongoing basis, his beard growth would return to normal, just as it would decline when he returned to the island after his weekend trysts. Given what we know about androgen receptors in the skin cells of the face, the surge in testosterone in anticipation of attracting the man's sexual partner most likely promoted beard growth as an outward sign of virility.

Clearly hair growth in men is far more complicated than a simple reflection of testosterone levels—in fact, concentrations of testosterone in the blood are roughly similar between bald and non-bald men. However, there is some difference in the levels of *sex hormone binding globulin* (SHBG), a major protein in our circulatory system to which most of the hormone molecules in the blood are attached.[65] Bald men have slightly lower levels of SHBG, allowing more testosterone to float freely in the blood and tissues and thus more widely exert its effects. To put it another way: while the total testosterone level is not lower in non-bald men, the *available* testosterone might be. Even more interesting is that SHBG levels may be influenced by outside factors, including some differences in environmental, dietary, or lifestyle conditions. For example, obesity may increase SHBG levels and therefore reduce the availability of "free" testosterone to do its thing.

Does that mean overweight men are more likely to have full heads of hair? Again, there are so many individual elements at work, it would be nearly impossible to say.

FALL IN THE FAMILY

My big brother and I enjoyed a fairly classic Midwestern boyhood together, and we are still close, even though I live three thousand miles away from where he is in Minneapolis. He was always bigger than me, but as an adult he has grown into something of a gentle giant, standing at six foot seven and weighing somewhere between 250 and 275 pounds. His appearance invites a bear hug, and his demeanor exudes approachability. Welcoming eyes and a quick, deferential smile adorn what I maintain is one of the largest heads in existence. We joke that it, like the Great Wall of China, can be seen from space.

Big heads are a family trait, and my brother, my late dad, and I all discovered just how big while playing high school football. I had the second-biggest helmet size on the team, my brother had the biggest on his, and in my dad's case they had to purchase a helmet from the local NFL team, the Washington Redskins, in order to outfit him. The detail that I find most interesting (and most relevant to this book), however, is not the way in which our heads are similar but the way in which they are different—specifically, the divergence of hair patterns they display.

My dad had classic male pattern baldness that began when he was a teen. When I was growing up, he was among the few bald dads in the neighborhood, and it was the subject of gentle ribbing at home, to which he responded with charming self-deprecation. When he and my brother joined a local Indian Guides group (a sort of Cub Scouts-meets-Survivor organization for men and their young sons), he adopted the moniker "Falling Hare." (Puns are another, more serious, family disease.) My brother became "Little Rabbit." My father's hair pattern was similar to his brothers', but my brother and I, now in our late forties, have developed completely, remarkably different hairlines. He is nearly as bald as our father was and makes semiweekly trips to the barber to maintain the shaved-scalp look. My own hair, however, has experienced only light thinning on top.

The conventional wisdom about baldness coming from your mother's father is *mostly* true. There is indeed such thing as a baldness gene. But given that my brother and I share a mother, why is my brother bald and I am not (yet)? It turns out there are actually *multiple* baldness genes. The interaction of these genes may be the key to understanding how baldness expresses itself (or not).

My brother and I both inherited our single X chromosomes from our mother. Perhaps surprisingly, the androgen receptor is encoded by a gene located on the X chromosome, which is to say that we not only inherit baldness from our mothers but we actually inherit our androgen receptors from our mothers, too. It's ironic, really, that the gene most important to virility—the one responsible for the androgen receptors that produce all the myriad "signs" of maleness—comes from Mom, not Dad. The scientific beauty of it is that there is another gene, the *SRY gene*, which exists on the Y chromosome that comes from the father, and its main feature is to turn on testosterone production and the genes that lead to the formation of male anatomy. Nature has an interesting way of balancing itself.

Androgen receptors are major drivers of baldness, which means that the gene controlling them determines much of our follicular fate. Current research also implicates a gene called EDA2R, which sits extremely close to the AR gene on the X chromosome.[66, 67] It contains the blueprint for a receptor found in hair follicles. When the gene works normally, it activates cellular substances that enhance the activity of androgen receptors, like a self-promoting feed-forward system. When an individual has a variation—called a *single-nucleotide polymorphism*, or SNP (pronounced "snip")—in one small segment of this gene, however, it alters the gene's ability to promote growth in the hair follicle, and this is another major driver of baldness.*

Consider the evolutionary significance of the proximity of EDA2R to the androgen-receptor gene. Throughout the passage of hundreds of thousands of generations, our chromosomes have undergone many changes, including breakage, the insertion of part of one chromosome into part of another, and even deletion of whole segments. This random movement and damage in our genome is one of the drivers of spontaneity and variation in nature. So with all that going on, two genes that remain located near each other on a chromosome are said to be in "linkage disequilibrium," implying that it is unlikely to be random that these two genes are near each other. The AR gene and EDA2R are two such linked genes.

Returning to the comparison between me and my big brother, we should have very similar X chromosomes, and therefore we probably have pretty much the same androgen receptors, and we might even share the same EDA2R gene. So why have our hair patterns diverged? Another recently identified "baldness gene," which may interact with EDA2R, likely provides the answer.

The full nature of this gene has not yet been fully characterized (for instance, we don't yet know *how* it confers baldness), but we do know its location on chromosome 20—20p11, which means it is in position 11 on the "short arm" of chromosome 20—and we think it might contain important information about male pattern baldness. This mysterious gene was discovered by researchers looking at full genome analyses of approximately five thousand men in Iceland, Switzerland, the Netherlands, and the United Kingdom who had donated genetic material and personal information about their appearance, behavior, and so

---

*The discovery of EDA2R is relatively new, so we need further research into where else this receptor shows up in the body and what kind of key role it plays in our survival, if any.

on. The researchers found a SNP on chromosome 20 that, when present in a certain pattern, seemed to increase the likelihood of baldness by 60 percent. However, the odds shot up to a *700 percent* increased likelihood when both the 20p11 gene variation and EDA2R variation were present in the same individual.[68] This 20p11 gene variation might explain why my hair pattern diverges from that of my older brother.

The fact that the second baldness gene was discovered in northern Europeans may also explain why male pattern baldness is about twice as common in men of European descent as in those of Asian descent.[69] It's even less common in men of African descent. Genetic anthropology studies suggest that this gene has been significantly affected by the migration of human populations away from our sites of mutual origin in Africa. One study showed that the differences in the pattern of EDA2R gene structure between East Asians and Africans is among the most divergent in the entire genome.[70] It is theorized that the divergence is so great because the gene variations developed *after* the migration of early humans out of Africa.

THE WHY OF BALDNESS

Does this tell us anything about the evolutionary "reason" behind baldness? I actually think it does.

At some point in time, this new mutation struck randomly in some prehistoric man on his way to the region that would become northern Europe. Perhaps this made his appearance different enough from the males around him that he became attractive to potential mates. Or perhaps something else was going on—after all, given the short life spans of earlier humans, and the fact that mate selection would likely have occurred young, his baldness wouldn't necessarily have been evident. Because baldness seems to involve increased androgen-receptor activity, *other* traits that also come along with this configuration of androgen receptors might have made our eventually-bald guy more attractive or simply bettered his chances of survival.

In a cartoonish way, I imagine two early human males standing somewhere at a fork in the road in the prehistoric Middle East. One has a lot of muscle but little hair, while the other has less muscle but more hair. Nearby, two early human females look on, contemplating their choice. Perhaps the one who coupled with the balding man went northwest and the one who coupled with the non-balding one went northeast. Note that this scenario would suggest that, in humans, the baldness story is not only about *natural* selection but *sexual* selection. Geneticists refer to this phenomenon as "positive selection," meaning that those who carry the gene are more likely to be selected by prospective mates, thereby essentially "selecting" the gene to be part of the greater population.

Reverse-engineering evolution is never simple, of course, but from what we can see about how the EDA2R gene has persisted over time—specifically, its continued proximity to the androgen receptor gene—the characteristic of baldness would have traveled alongside other traits controlled by the activity of androgen receptors: your traditional "survival" traits like strength, hunt skill, and dominance. Put simply, it's possible that baldness may go along with being a stable provider, or at least it did long ago, and as a result it has been preserved in our genetic code.

THE BUSINESS OF BALDNESS AND THE DANGERS OF BALDNESS CORRECTION

Whatever its evolutionary provenance, baldness isn't traditionally considered to be an attractive feature, and the amount of money spent annually on hair replacement worldwide is estimated to soon exceed the amounts spent on AIDS research. Survey data from the International

Society of Hair Restoration (ISHR) states, albeit self-servingly, that 47 percent of balding men would spend their life savings to get back their full head of hair. Of those surveyed, 60 percent say they would rather have more hair than money or friends, and 30 percent would give up sex if it meant they would get their hair back. Even taking into account the selection bias inherent in collecting data from people who fill out surveys from a hair-restoration company, those are still significant numbers.[71] According to the ISHR, $2 billion is spent every year on surgical hair restoration, a fact brought to the fore in 2013, when Bill Gates derided this figure in the media, pointing out it was double the amount spent on controlling malaria, a disease that affects more than 200 million people worldwide and kills approximately 700,000 every year.[72] Whether you agree with it or not, fighting baldness is a big business, and one not likely to go away anytime soon.

A few years ago, I began fretting over the beginnings of a bald spot, so I started using Nioxin, an over-the-counter shampoo, and for a while I catalogued the stability of my mild bald spot with iPhone pictures after every haircut. (My longtime barber, Robert, was complicit in this neurotic nerdicism.) Eventually the minor thinness stabilized, and I continue to swear by the shampoo and have even recommended it to numerous friends. Would it surprise you to know that in using the special shampoo I am manipulating my testosterone—at least as it exists on my scalp? The formula is essentially a detergent with a special affinity for steroid molecules, meaning it washes the testosterone and DHT (another androgen hormone) right out of my hair follicles. There are a number of techniques for removing testosterone and DHT from the hair follicle—but they may not all be completely safe.

While being bald in today's society may have its downsides (studies show women do tend to prefer men with hair), baldness correction can have its own harsh consequences. In fact, a touch of baldness forever altered the life of an acquaintance of mine named Kevin. At age thirty-seven, he had a hair-transplant procedure at a local clinic. His doctor suggested that taking a medicine in addition to the transplant might produce even better results.

"I couldn't fill that prescription fast enough," he told me.

The drug was Propecia (generic name *finasteride*), which is also sold as Proscar, the differences between the two being a matter of quantity: Propecia is the 1 milligram formulation prescribed to promote hair growth, while Proscar is the 5 mg tablet prescribed to shrink enlarging prostates. Finasteride works by blocking the enzyme *5-alpha reductase* (5AR), which is responsible for the conversion of testosterone into the even more potent androgen *dihydrotestosterone* (DHT). The more testosterone turned into DHT courtesy of 5AR, the greater the effect on androgen receptors. In order to stretch his dollar a bit further, Kevin got a prescription for the 5 mg dosage and quartered the pills to approximate the dosage he was supposed to take. Simple economy of scale.

Kevin's confidence boosted after the transplant, and he believed it was well worth his time and money, even for a family on a tight budget. Yet, it didn't take long before he knew that something was a little off.

"About six or seven months after I started taking the pills, I said to myself, 'Man, Lisa and I haven't had sex in a while.'"

He couldn't even remember the last time he had made love to his wife. He told me it just hadn't crossed his mind. He had forgotten that sex was something he did, that he liked and wanted.

"I wondered what in the world was going on," Kevin said.

The strangest part was that he was now thinking about sex only intellectually, as a thing that is *done*. He wasn't even *feeling* the loss of his libido; he just noticed that it was gone, like a neighbor you used to wave to every day and suddenly realize you haven't seen in a while. You don't necessarily miss him, you just notice his absence.

Up late at night, Kevin scoured the internet for an answer to his loss of libido, and, consistently, the medical websites stated that "some medications" might be to blame. Bingo; it was the finasteride. He went off the medicine and figured that would be the end of it. Or the *return of it*, if you will. But his libido did not come back; in fact, the situation seemed to get worse.

Over the next year Kevin continued to descend into a spiral of insomnia, lack of interest in work, and fatigue. Doctors call this *anhedonia*—an inability to experience pleasure, not just sexual pleasure but pleasure of any kind, and it's a pretty miserable state to be in.

He started on an antidepressant, and was then prescribed testosterone gel to boost his testosterone levels. He took it for a month or two but decided not to refill the prescription, having been told (likely correctly) that taking the drug could lead to a lifelong dependence on testosterone supplements. He decided to tough it out without supplementation and hope things would improve over time. And they have—although very, very slowly. It's been seven years.

"Now, I just want a simple life. I stepped down from my supervisor job because I don't want the stress of being responsible for other people. I just want to go to work, get it done, and come home."

Kevin's convinced that this loss of ambition is another side effect of having taken finasteride.

"I don't feel any excitement at all," he tells me. "And that includes, or should I say *especially* includes, in the bedroom. It's like there's a disconnect between my brain and my penis."

For Kevin this means his libido is nil and his ability to orgasm is gone. He can get an erection, with pharmacologic help, but sex is now a chore, something he does to keep his wife happy. And while he's grateful to have a supportive wife and a loving family, his depression casts a shadow over everything.

Another person I spoke to, Lars, was even less fortunate. Approaching forty, he was married to his high school sweetheart, father to a newborn baby, in a solid job with an up-and-coming biotechnology company, and raising horses on the family ranch in Southern California. But as with Kevin, the problem of his thinning hair nibbled away at him. He, too, sought the advice of a hair-transplant surgeon, who suggested Lars first start with Propecia, as it was the least invasive treatment and had the lowest risk of negative side effects. Besides, Lars's baldness wasn't yet severe.

Within a few weeks of starting the drug, Lars was finding it harder to focus on his work and was prone to episodes of mood swings and crying.

"I started going for long walks during my lunch hour to see if I could shake myself out of the funk. Instead, I spent most of the walk just *weeping*. I just didn't know what was going on with me."

His marriage crumbled under the strain, and his wife, Laura, left him.

Eventually he turned to alcohol, in part because an internet posting he came across suggested that *allopregnanolone*, a hormone in the brain that is low in depressed individuals, might go up with alcohol consumption.* He was self-medicating.

In fact, Lars's blood testosterone level was low—just a little under 200 nanograms per deciliter†—and he started on testosterone supplementation under the care of a sexual-medicine specialist. He also hit the gym like a madman, pumping iron and putting on muscle as another way to distract himself from a life he felt was slipping away.

Unfortunately, this only made things worse.

Lars's testosterone level zoomed up, from slightly below normal right through the normal range and beyond. At one point it was all the way up to 1800—more than twice the highest point of the normal range. He was struggling to get along with his now ex-wife, desperately wanting to continue being an involved dad. On Laura's birthday, he saw an

---

*Interestingly, in rats that get prenatal finasteride, alcohol consumption goes down. Data has also been published that shows men with post-finasteride syndrome (PFS) consume less alcohol than would normally be expected. Men on anabolic steroids for bodybuilding also consume less alcohol than bodybuilders not on steroids. Blocking testosterone's effect on the fetal brain, it seems, reduces craving for alcohol later in life.

†The normal range is about 250 to 700; at his age you would expect it to be between 500 and 650.

opportunity to be together with the family and showed up bearing gifts at the home of Laura's mother, where she and the kids were celebrating.

His veins coursing with testosterone, he was on a hair trigger, and the first irritant was his ex-mother-in-law's big Doberman, barking angrily. Although it didn't do anything except bark, Lars grabbed the dog by the scruff and erupted, "Control your *fucking* dog!"

"Then they ganged up on me," he recalled. "And I said some stupid things. She—she just made me really angry."

Laura called 911.

Now, Lars can see his kids only for six hours on Sundays and he can't talk to his ex at all. There's no hope for reconciliation; she's moved on. He lives in the country these days and pretty much keeps to himself. Propecia is a low-dose drug, and it was supposed to manipulate testosterone only in the scalp. Instead, it had done a number on Lars's brain, bringing on episodes of depression, irritability, and aggression that changed his life.

Or had it?

It's easy to be skeptical of these stories, and I was when I first heard them. Maybe the fates of Kevin and Lars had nothing to do with the medication they took for hair loss. Maybe they were already prone to depression; perhaps the fact that they were bothered enough by their hair loss to undergo surgery suggests they had low self-esteem to begin with. There are plenty of possibilities, but I'm inclined to believe that the medication had something to do with it, for several reasons.

First, we're not talking about just a handful of men who experienced similar symptoms in response to taking finasteride products—it's many hundreds, maybe more. A 2012 paper in the *Journal of Clinical Psychiatry* reported that a whopping 44 percent of former finasteride users who experienced sexual side effects also reported thoughts of suicide.[73] Yikes.

Second, it's biologically plausible that finasteride was causing these problems because we know that 5AR, the enzyme that converts testosterone into DHT—the enzyme blocked by finasteride—is heavily active in the parts of the brain known as the "reward pathways." As we saw in the previous chapter, impairment of these circuits can lead to lack of enthusiasm for things like sex and work.*[74]

Third, DHT and testosterone aren't the only hormones that interact with 5AR, and in fact research into the dynamics of various hormone levels in cerebrospinal fluid has revealed that they decrease in response to finasteride or other drugs that impair 5AR. Many of these hormones play a role in the regulation of a person's mood.

The term *neurosteroid* refers to a steroid synthesized in the brain. It was coined in 1981, to describe the discovery of high levels of DHEAS—a "raw material" hormone that works as a building block for other hormones, including testosterone—in the brain fluid of a patient following the removal of his testicles. That this hormone so closely related to testosterone had been produced not by the testicles but by the brain was a major revelation.

Neurosteroids like DHEAS and others interact not only with androgen receptors but also with the two types of receptors for the chemical known as GABA. *GABA (gamma-Aminobutyric acid)* is found throughout the brain and, like testosterone, interacts with its receptors—GABA(a) and GABA(b)—to exert its function. Activation of the GABA(a) receptor has sedating and calming properties, and research has shown that individuals with anxiety disorders have an underactive GABA system. Drugs like diazepam (Valium), ethanol alcohol (the kind of alcohol in booze), and barbiturates have what are called *GABAergic* functions, meaning they act like GABA to activate GABA(a), reducing anxiety and inducing calm. A neurosteroid called *allopregnanalone* also has GABAergic activity, and research has shown

---

*Just as the androgen receptors in the scalp are subject to genetic variation, so they are in the brain. Individuals can have differing levels of sensitivity to finasteride (as they do to testosterone) based on variations in and combinations of androgen receptors and their 5AR subtypes.

not only that levels are low in clinically depressed people and temporarily low in women during the pre-menstrual part of their cycle,* but that the level also dropped in patients on finasteride. So, there is the chain of circumstantial chemical evidence that finasteride may indeed be responsible for the mood effects suffered by many men who used it for hair regrowth.

The official name for the group of symptoms Kevin and Lars experienced is *post-finasteride syndrome* (PFS), and perhaps one of the most troubling aspects was that it seemed so permanent, with the effect often persisting *years* after discontinuation of the drug. One clue to why that might be comes from animal studies, which in this case also tell us more about the feed-forward relationship between testosterone and the androgen receptor. A group of Italian researchers gave finasteride to rats and noticed that the number of androgen receptors in their brains went *up*.[75] Moreover, the effects persisted long after the drug had been discontinued. To follow up on this finding, they then called in men with PFS, took skin from the penis, and found that the density of androgen receptors in men with PFS was about *twice* that of those without.[76] Now, remember the idea of the testosterone bell curve and damping effects (little testosterone, little growth, more testosterone, more growth, even more testosterone, *reduced* growth)? I think this is what we are seeing here. With a greater concentration of receptors, the organ becomes more sensitive to testosterone and at a certain point, paradoxically, that sensitivity may shut down.† The Italian researchers

---

*Drugs that increase or mimic the activity of allopregnanolone are currently being developed to treat conditions that are the result of hyperactive brain circuits, like epilepsy and post-traumatic stress disorder.

†The androgen receptor's activity is driven not only by testosterone and by how many receptors there are, but also by the balance of other molecules that bind to it and activate it (co-activators) or repress it (co-repressors). The recognition of a shift in this balance as the AR amplifies in prostate cancer cells enabled the development of Enzalutamide, a very effective prostate-cancer drug.

have subsequently demonstrated that patients with either very short or very long CAG repeats in the androgen receptor suffer these effects,[77] reflecting the complexity of this chemistry as well as the plausibility of this mechanism.

Did Lars's brain adapt to the blockade of 5AR by amplifying androgen receptor levels and, in turn, its sensitivity to testosterone? It's possible. It does look as if, in response to the 5AR blockade of finasteride, the brain produces more receptors in order to become sensitive to lower levels of the hormones. This is a familiar concept to those of us who treat prostate cancer, for this is the same adaptation that cancer makes to survive after testosterone is wiped away. As we've mentioned in previous chapters, it is possible that having greater androgen-receptor activity could make you not only more sensitive to testosterone—for instance, causing levels to skyrocket, as they did with Lars—but also more sensitive to its absence.

All this said, there are plenty of men out there who swear by the drug. Why does it work fine for some and not others? It seems possible that something about the subtypes of 5AR in the brains of these men is responsible for the variability in their sensitivity to inhibition with finasteride. I am also struck by the fact that this syndrome tends to manifest itself in young men who are taking *low-dose* finasteride (1 mg) and not older men who take the higher dose of 5 mg to combat prostate enlargement. Perhaps the changes that occur with age, as discussed in the previous chapter, have something to do with this, or perhaps it is another bell curve.

We've seen throughout the book that anytime you manipulate a piece of the testosterone system, you risk unexpected changes. The important difference, to me, is that in the case of using finasteride for baldness, this risk is not a necessary trade-off to treating a deadly disease like prostate cancer; it's the cost of treating a receding hairline.

SO, WHAT DOES BALDNESS REALLY SAY?

We still haven't answered the question of why so many men go to such great lengths to combat hair loss in the first place. You might imagine that if hair loss were truly a marker of unattractiveness, it would be disappearing over the generations, as balding males would have trouble mating and the gene would have been removed from the gene pool. The fact that it still affects more than 40 percent of adult men clearly argues against this.* Even though baldness typically occurs after the age at which a man would have reproduced, that age of course has risen, and even baldness that strikes earlier remains fairly common. Baldness itself may not have anything to do with sexual selection at all. What does baldness really tell us, if anything?

Some research suggests it telegraphs health problems. It is in fact true that baldness may signal a higher risk of serious health issues, including diabetes, heart attacks, and aggressive prostate cancer.[78] Such relationships first came to light in the early 1990s, when researchers at Boston University assessed the hair of 665 men under age fifty-five who were admitted with non-fatal heart attacks to a cluster of New England hospitals. They reported some degree of baldness in 34 percent of the heart-attack victims and severe baldness in another 24 percent of the subjects. As a control arm for this study, they evaluated the hair of over 770 men of the same age admitted to the same hospitals

---

*Several years ago the British tabloids took quite an interest in the fact that, at the time of his engagement and subsequent wedding to Kate Middleton, Prince William displayed a receding hairline. He was twenty-eight. Some speculated he was in a hurry to tie the knot because he was in the early stages of baldness and, should that continue, he might have difficulty finding a mate. Can you imagine? The heir to the British throne? Trouble finding a mate? One doesn't need a PhD in history to know that, through it all, the royal families of the British Empire have managed to find mates with relative ease, and if you know anything about the famously bearded Henry VIII, I'd say that finding *too many* mates may have been more of a problem.

but for medical problems other than heart attack. In making these comparisons, they revealed that the relative risk of a heart attack for those men with vertex (top-of-the-head) baldness was 3.4. Translation: bald men were almost three and a half times more likely to experience heart problems.[79]

Interestingly, this original study was done during a time when topical *minoxidil* was gaining popularity as a baldness treatment, and there had been concern that the drug itself was causing heart problems. The Boston University study was in fact initiated to determine whether there was a link between minoxidil and heart attacks. It turned out, however, that *baldness* was the link, not the baldness treatment. (This is an important observation in light of the finasteride data; there also we must ask whether the problem is the treatment or the disease itself.) In 2013, a Taiwanese research group reported a twofold increase in the risk of death from diabetes in men and women with androgenic alopecia, and in fact the severity of hair loss correlated with the risk of death. Moreover, in a multivariate analysis, baldness emerged as an independent risk factor for the disease.[80]

There is actually a silver lining here. Although men may not be ready to embrace baldness, perhaps they can at least use baldness to their advantage: it may alert some individuals to their risks of heart disease, diabetes, and prostate cancer. Baldness is not a disease or even a medical problem per se, yet it seems to be an early warning sign that other problems may loom in the future. Balding men especially, this data suggests, should pay particular attention to blood pressure, cholesterol, and blood-sugar levels. In a recent study on prostate cancer, moderate baldness (meaning baldness only on the vertex) at forty-five predicted prostate cancer at age seventy.[78] I wrote an editorial on the subject for an oncology journal and worked to provide a positive perspective, stating that the baldness gave these men a twenty-five-year warning. We may not have a cure for baldness yet, but we *can* prevent other afflictions, including heart attacks, which are pretty effectively avoided by controlling cholesterol and blood pressure, taking daily aspirin, and making other lifestyle modifications. We can even cure prostate cancer if we detect it early enough. Primary care doctors may wish to start considering what baldness can reveal about their middle-aged male patients. Keep in mind that these associations aren't casual or accidental; both prostate cancer and baldness (and maybe even heart disease) are conditions associated with chronic persistent exposure to testosterone.

Hair loss is all around us. Some days it seems more the norm than the exception. And, in my unscientific data collection on the subject, it seems that the men who worry about their baldness fret about it far more than do the women who love them. At this stage in human evolution, baldness might even help men attract mates. One study from several decades ago explored the ways in which various hair patterns (clean-shaven with a full head of hair, bald with a beard, bald and clean-shaven, bearded with a full head of hair, etc.) were perceived. The subjects were asked to rate the men in the photos in the categories of aggressiveness, appeasement (willingness to compromise), social maturity, and attractiveness, and the results suggest that balding men are often seen as having greater social maturity, less aggressiveness, and more willingness to compromise than subjects with full heads of hair.[81] In our modern civilized and predator-free society, this is good news. The receding hairline may present the image of a better and more stable husband and father; now that mate choice is no longer strictly about survival—when was the last time you needed to hunt for food or fight off a saber-tooth tiger?—these softer personality traits may be becoming more desirable, and that's probably a good thing. Maybe an acceptance of baldness as a marker of long-standing and *mature* virility is the way to go. After he reads this, I suspect that my brother will think he's more of a man than I am because of it. He may be right.

Chapter Eight

THE DARKEST DEMON: THE BIOLOGICAL UNDERPINNINGS OF SEXUAL AGGRESSION

On a cool December night in 2012, in the Palam district of New Delhi, India, Jyoti Singh, a twenty-three-year-old physical therapy student, stepped out to see *The Life of Pi* with her male friend Avnindra Pandey, a software engineer. The movie ended at 8 PM and they made their way to the nearby Munirka bus stand, where a private bus was idling. A young boy leaned out the front door, smiled, and urged them in. Five others were already inside. As they climbed aboard and the boy asked for payment of a twenty-rupee fare, Avnindra scanned the others on the bus and instantly felt uneasy.

The bus lurched forward and the driver pulled the door lock. The lights dimmed. Three of the men abruptly rose out of their seats and moved toward Avnindra. The first man punched him cold in the face and they scuffled. Avnindra managed to break the glass of a partition door as Jyoti screamed in terror and reached for her phone to call for help. Two of the men pulled her back and took the phone. Avnindra was dragged to one end of the bus and beaten with a metal rod. Jyoti was dragged by her hair to the other end.

Then, and for the next eighty-four minutes, the six men took turns raping her. The violation included a foreign object that was inserted with such force that her uterus was mutilated and she was partially disemboweled.

When done, they dumped Avnindra and Jyoti out onto the street—barely clothed, bleeding, torn apart, and semiconscious. Several cars slowed or even stopped, but no one offered help. Eventually a police car approached, and then an ambulance, and the couple was taken to the hospital. Avnindra eventually made a full recovery, but Jyoti, after a series of surgeries to repair her injuries, died, thirteen days after the attack. All six men were apprehended, and the four surviving adult perpetrators were sentenced to death.*

More than 33,000 rapes were reported in 2013 in India, the year after the attack on Jyoti Singh, and it is likely most rapes are not reported to authorities. India has had a particular problem in this area, but rape is a worldwide issue. The US Department of Justice estimates that only about 35 percent of rapes in the United States are reported, and their statistics suggest that in any given year, about one in a thousand individuals will be the victim of rape.[82] Criminal science, psychology, and biology all come together in the study of sexual aggression; a true understanding of the phenomenon, if possible, will be obtained only through an appreciation of how these systems interact. I can't hope to untangle all of these threads here, but I can explore how hormones, and in particular testosterone, might play a part.

---

*The juvenile was convicted of rape and murder and given the maximum sentence of three years of imprisonment in a reform facility. The fifth adult died in jail, possibly of suicide, three months after he was arrested.

A BIOLOGICAL BASIS

Rape is a worldwide human phenomenon. It has occurred in every society during every era of history. The outlawing of rape in one form or another is also something of a human universal, dating to the earliest records we have of codes of law and religious doctrine.

Evolutionary biology tells us that, at its most base, heterosexual rape is an act of biological and genetic *economy*—not a word you were expecting, I would wager. Speaking strictly from the perspective of evolutionary biology, a biologist would say rape has been imprinted in societies as competition over, and the theft of, a scarce resource: the reproductive potential of the female. This does not come close to explaining the complex implications of rape in psychological and sociological terms, but *biologically* we can say that this is how sexual violence got into our genome, so to speak, and why it has persisted over the generations.

To understand why female reproduction is considered a scarce resource, consider the huge disparity between male and female contributions to the process: a human male can contribute to the reproduction of the species every day (maybe even more than once per day, if he has multiple partners), as long as he is provided the opportunity to do so. With every ejaculation, he releases 250 *million* sperm, all hoping to find a single available egg. A human female, on the other hand, can only "contribute" once every nine months at the most. And when she is not pregnant, she is only fertile for a few days each month. The biological clocks of the sexes, if you will, operate on vastly different cycles. This is generally true in all of the more complex organisms, and indeed rape is not an entirely human construct; it has been observed in gorillas, orangutans, and chimpanzees—other species limited by long gestation times.

Another set of data in support of this theory is that throughout history an estimated 80 percent of men have reproduced successfully, compared with only about 40 percent of women.[83] Individual men in positions of

dominance have also not only sought but ensured their repeated reproduction with many different women through the creation of harems and
other polygamist structures. This itself is a form of biological dominance
in a male-to-male competition. It will not be surprising to hear, then, that
some genetic correlates of the virility triad are associated with sexual
violence—and with violence and aggression in general.

The associations are modest, but the correlation between hormone
levels and sexual violence follows the pattern you might expect. Higher
fetal testosterone exposure is associated with criminal behavior in general: a lower 2D:4D ratio (which signals more fetal testosterone) correlates to a higher likelihood that an individual will have committed a
crime. In fact, 2D:4D ratio has a stronger association with crime than
does age or race. It is an independent predictor.[84]

A few years before the gang rape of Jyoti Singh, researchers in
Hyderabad, India, analyzed androgen receptor CAG length in a group
of prisoners and drew associations between that data and the crimes
that had led to their subjects' convictions. Their sample size was 645
subjects, of which 241 had been imprisoned for rape, 107 for murder,
and 26 for both rape and murder, with 271 men from the same region
of India without a criminal record used as controls.

The average number of CAG repeats in the AR gene (remember:
the shorter the CAG repeat, the faster the AR, and the more sensitive
it is to testosterone) is about 21. The average repeat was 18.4 for the
rapists, 17.59 for the murderers, and 17.31 in those who raped and
then murdered their victims. Almost 80 percent of the rapists had a
CAG repeat length less than 21, whereas only about 40 percent of the
control subjects did. These findings met standard statistical criteria for
significance—meaning the results were less than 1 percent likely to be
observed by chance—and on the basis of this data the authors concluded that the rapists had more-active androgen receptors.[85]

Although this data is provocative, I view it with some skepticism.
While I don't question the findings, a couple of clarifying points can

provide some important perspective. First, the statistical methods here are *univariate,* meaning they don't allow for correction for other factors. In this case they did not correct for things like prior abuse, psychiatric diagnoses of the inmates, or socioeconomic status. Second, and most obviously, while the data clearly points to the low CAG repeat as a biological type associated with rape, it isn't making the point that this finding *is specific to* rape. The data merely suggests—and this is the best we can do—that this single genetic factor in men *may be an augmenting factor* in a complex set of psychological, emotional, and even societal factors that culminate in this act.

Lest we think this data is unique to India's particular genetic pool, or that it is unique to rape, researchers in China found a similar relationship in that country's general criminal population. They found that 7.5 percent of criminals had CAG repeats measuring 17 or below, compared with only about 2 percent of non-criminal control subjects. The Chinese data suggested less specificity for the crime of rape and suggested more generally, as have other studies, that a short CAG repeat length is associated with aggression as a whole.[86] Again, this data is provocative, but what of the 92.5 percent of incarcerated criminals who did not have short CAG repeats? Thus the finding lacks much in the way of both sensitivity (that is, the number of criminals with low CAG repeats is not that high) and specificity (meaning there are other effects besides criminal behavior that could arise from a low CAG number).

And yet to some degree this data *does* place testosterone and the androgen receptor at the scene of these crimes, so to speak. Prior work had shown that serum-testosterone levels were not higher in rapists or child molesters; however, it was found that levels of DHT (the more potent molecule into which testosterone is converted) *are* higher in some aggressive people. Clearly the interaction between testosterone and criminality is anything but straightforward.

Modest as the existing data is, let's assume it's significant. A number of perspectives can be taken in analyzing the findings. First, the evolutionary perspective. Of course, a pregnancy resulting from rape—and rape results in pregnancy about 6 percent of the time (versus 3 percent with consensual sex)[87]—allows the rapist's genes to be passed on. An important detail to note here, however, is that although evolution may have selected for the preservation of genes that contribute to a person's likelihood to commit violent crimes, this selection wasn't specific for sexual aggression, it was simply selection for aggression in general, which, as we have seen, is beneficial in other ways. This mirrors what has been called the evolutionary neuroandrogenic (ENA) theory,[88] which says females seek to mate with males who exhibit a significant potential for provisioning—aka the ability to provide for them and their subsequent children, something we discussed in the first chapter of this book. Aggression is among the features that can augment a man's success in this area (think of hunting for food specifically), and as the successful provider is more likely to win over a mate, we can say, based on the ENA theory, that male aggression is evolutionarily connected to the need to compete with other men in order to win the scarce resource that is female reproduction. In cases where the aggression meant to facilitate competition swings to the extreme, rape and other violent behavior may result.

SEXUAL SELECTION AND THE ISSUE OF CONSENT

One of the founding pillars of the ENA theory—which extends to other animals besides humans—is the *consent* of the female. Per the tenets of Darwin's theory of sexual selection, she is in charge, since she is the one choosing among many potential mates. Lack of consent is, in fact, what makes rape rape, and although the concept has a clear definition, it is not as easy to delineate in some real-life situations. The consent issue has been the subject of many recent debates, and one of the most common arguments is over how substance abuse affects a person's ability to consent. If, for instance, a woman is heavily intoxicated when she agrees to sexual intercourse, is it fair to say she consented? Does any-thing change if the man is also heavily intoxicated? How should society prosecute the crime if the mental capacity to consent exists as a gradi-ent influenced by alcohol, drugs, or other factors? One shocking point of view that came to light in the context of the abortion debate was the idea that women's bodies can somehow protect themselves from the biological consequences of rape (i.e., pregnancy). Recall the 2012 com-ments by Missouri Senate Republican candidate Todd Akin, who said in an on-air interview, "From what I understand from doctors, [preg-nancy from rape] is really rare. If it's a legitimate rape, the female body has ways to try to shut that whole thing down."[89]

Legitimate rape? I wince as I write it.

He is wrong, of course, but is there evidence that such a thing might be possible? As it turns out, there is—although not in humans. Maybe, just maybe, Akin was confusing what he had read about ducks with what he had "heard from doctors." Stay with me here. There is in fact a biological mechanism through which female ducks can "shut that whole thing down."

Birds in general are a useful model for the study of the relationship between androgens and aggressiveness because we have a great deal of data on the relationship between their hormones and their behav-ior. For example, seasonal variation in hormone levels has been stud-ied in many bird species: in times of environmental stress, starvation, and restriction (e.g., the winter) the production of sex hormones shuts down, but it returns with refeeding in the springtime, often accompa-nied by an increase in aggression as the birds compete for territory and mates. Forced copulation is the norm in most bird species, including ducks much of the time—but not always. The medium-sized mallard, like the ones you see in parks all over North America, has a penis that is about 40 centimeters long. It's actually longer than the rest of the duck! In addition to being very long, it is coiled like a corkscrew in a counterclockwise direction. The long duck penis is the evolutionary counterpart to the highly defensive duck vagina, which is also coiled like a corkscrew in a counterclockwise direction. Within the turns of the vaginal corkscrew are several tiny pockets, like *culs de sac*. If a penis enters a pocket instead of penetrating all the way into the vagina, it will not get close enough for the sperm to fertilize any eggs. And there's more: the end of the vagina curls *clockwise*, against the direction of the penis—that is unless the female duck's body *consents* to allow it to enter completely. When a female duck approves the copulation in accordance with the desirability of the mate, her vagina actually relaxes, allowing successful copulation, and hopefully fertilization, to occur.

This is an interesting little trick of nature, but is it about consent, or about defense? Perhaps a little bit of both? And what does this have to do with testosterone?

A group of scientists in Sweden set out to test the idea that female ducks are more likely to reproduce with males who have higher testosterone and are therefore perceived as "fitter" mates (Darwin's term). They confined two groups of male ducks and gave one group unlimited food while the other was put on what was basically a portion-restricted diet. Over time, testosterone levels rose in the well-fed ducks compared with their underfed counterparts. The researchers then studied the likelihood with which female ducks would "initiate" sexual activity with males from each group. The ducks with an unlimited food supply had higher levels of testosterone—as well as dihydrotestosterone, its more potent form. They also exhibited more "male social display." Perhaps not surprisingly, the female ducks chose the better-fed ducks with the higher testosterone levels.[90] It is hard, in experiments like this, to separate the relative effects of the general health of the duck from the testosterone levels specifically, but in this context, testosterone levels were a sign of better general health.*

---

*If I had been consulted regarding this experiment, I would have recommended use of a classic 2x2 design. I would take poorly fed ducks and give them testosterone

Interpreting this science and its possible implications in humans goes a bit deeper than just noting that better-fed ducks breed more. Why do better-fed ducks have higher testosterone in the first place? This question is addressed by what anthropologists call the "challenge hypothesis," which, in turn, is a component of what is called "life history theory." This theory proposes that testosterone levels can be affected by life choices and challenges. That's what we see happening when testosterone levels rise in ducks that are not challenged by food scarcity. You may remember some of these concepts from Chapter 1, and our discussion of the winner effect.

Many modern humans aren't really "in nature" at all anymore, and as such we have managed to eliminate the physiological stresses that nature put on us. This fact is borne out in studies of men living in modernized, food-rich Western countries compared with subsistence farmers or hunter-gatherers living in remote regions of Africa. The near-complete lack of physiological stress (e.g., starvation, chronic infection) in the average Western male has created a situation in which, it has been suggested, men are living at the upper limits of their testosterone capacity. This situation is cited as one of the reasons why prostate cancer is more common in men living in overfed Western countries. Evolution has brought us to the point, through better nutrition and sanitation, where average testosterone levels are probably at their highest point *ever*. With so much of the hormone in play and so many of its original outlets out of the picture, it's easy to see how, given a particular set of circumstances, it might explode into sexual aggression and violent crime.

(to separate the feeding effect from the testosterone effect) and similarly would take well-fed ducks and deprive them of testosterone. However, I received no call from the Swedish ornithology community, probably because I was still in college at the time.

ANGRY YOUNG MEN

Of course, just because a man may be more biologically inclined to commit such acts doesn't mean he is unable to make different choices. That
said, there is evidence that young men in particular might be impaired
when it comes to controlling those urges, and in fact a report from the
federal Bureau of Justice Statistics (BJS) shows that young men enter
federal and state prison at higher numbers than older men and that
the likelihood of going to prison for the first time decreases with age.
Let's explore the workings of testosterone during puberty and at the
beginning of adulthood, as this is the age when impulsivity is highest
and young men are at the greatest risk of dangerous and even criminal
behavior. According to the BJS study, men are eight times more likely
to go to prison than women, and according to mortality statistics from
the Centers for Disease Control, the risk of death for a nineteen-year-
old man is three times that of a nineteen-year-old woman.[91]

It will not come as a surprise to anyone who has ever been around
adolescent boys and young adult men that these years are a time of
incredible change and growth, as well as a degree of . . . stupidity.
Around the age of fourteen or fifteen, the male brain undergoes a
stunning amount of development and at an amazing rate; whereas
in adolescent girls, the brain grows, physically, at a more gradual rate
throughout adolescence, in boys it's a steep upslope. If you think testosterone might have something to do with this difference, you're on
the right track.

But what does that actually mean? Does it mean larger brains
make boys smarter than girls? Nope. In fact, having more testosterone
and more active androgen receptors during adolescence may actually
have a *negative* impact on brain function. Researchers in Nottingham,
England, conducted a battery of tests on teenage male subjects and
found a significantly higher rate of depression and suicidal thinking
in boys with a combination of higher testosterone (than their male

peers) and shorter CAG repeats—the hallmarks of a more active virility triad.[92]

Neuroscientists hypothesize that the impulsivity, increased risk of
accidents, and proclivity to violence we see in young men may be due to
a mismatch in the maturation of two distinct brain areas. Specifically,
the parts of the brain related to impulse control and planning actually
mature later than do the parts related to reward and aggression. If you
look at the brain as a kind of a circuit board, it's as if the aggression
circuit and reward circuit are turned on but the impulse-dampening
system is not yet developed and therefore can't regulate the energy
flowing through the other channels. In this way, the adolescent male
brain may produce feelings of reward in response to risk-taking, and
have a harder time resisting impulses that fly in the face of good reason. Scientists at the National Institute of Mental Health set out to
measure this phenomenon in growing boys and girls, and, in effect,
to create a "movie" of the teen brain. What they found supports the
idea that the exact regions of the frontal lobe associated with impulse
control are the *last* to fully mature.[93]

I have always found it interesting that police and doctors perform
forensic autopsies on people who are killed while committing acts of
violence—such as the teen boys who commit the now-all-too-common
school shootings. Are the scientists looking for some biological clue as
to why teen boys commit mass killings? If they find something, does
that give us hope of preventing such tragedies in the future?*

---

*In 1966, twenty-five-year-old Charles Whitman murdered his mother, his wife,
and a number of random students on the University of Texas at Austin campus
before being shot and killed by police. His autopsy found a tumor in his amygdala,
the part of the brain that affects emotional regulation and is also well known to
contain androgen receptors and be affected by testosterone. In one of several suicide notes, Whitman himself requested an autopsy be performed on his body, as
he suspected there might have been some biological cause for his behavior. Some
have speculated that similar factors may be behind the mass killing in Las Vegas
in September 2017.

THE LINK BETWEEN TESTOSTERONE
AND SOCIETAL AGGRESSION

Earlier this decade a team of Russian ethnologists traveled to Tanzania to study the link between hormones and aggression in two tribal populations there: the Datoga, a tribe known for its aggression and a penchant for a polygamous lifestyle, and the Hadza, a pastoral, generally more peace-loving, monogamous group. Members of each group completed surveys and were interviewed about their aggression history (fights, etc.) as well as their reproductive history.

The results put the virility triad and its interwoven parts in context. Datoga men fathered more children, had higher aggression scores and testosterone levels, and shorter CAG repeats. As we have seen elsewhere, more fuel (testosterone) put into a more active engine (more-active ARs that result from shorter CAG repeats) added up to more aggression. Hadza men, by comparison, fathered fewer children and had lower testosterone levels and longer CAG repeats (less-active androgen receptors).[94]

What can this study tell us about aggression in groups, or even in entire populations? Would it be fair to say some humans are more "warlike"? Here again, we must remember that people are capable of moral judgment, and it would be too simplistic to say that aggression and violence are the products of some evolutionary tide against which we have no recourse. So, what makes the difference in whether we follow our dangerous impulses or choose another path? If testosterone is even partially to blame, is it because of the power of the hormone itself, or is it because testosterone can somehow disable what we call the moral compass? Are people more likely to commit acts of violence in groups, as with the men on the New Delhi bus? What does it tell us about war, perhaps the ultimate expression of violent aggression?

Again, let us turn to the animal kingdom. Mangrove rivulus fish live in rivers in North and South America and look like a cross between a drab-colored aquarium fish and the minnows I used to catch in a net off our dock in Wisconsin. A particularly hostile species, they are a good experimental model for the study of animal aggression. If placed in proximity to one another, male rivulus fish will inevitably fight. In the fish who repeatedly lose, we see reduced levels of androgen receptors in their brains, and, as you might expect, we see the opposite effect in repeated victors.[2] This is yet another example of the self-perpetuating feed-forward nature of testosterone and its receptor; essentially, the more aggressive fish are benefitting from the winner effect.

In this case, one of the most interesting phenomena was that the fish with low testosterone levels at the outset actually saw the greatest increases in brain AR levels, almost as if compensating for their low levels of testosterone. This is similar to what we have seen in previous chapters. These examples teach us something about the primacy of the testosterone/AR relationship: the pathway works to preserve testosterone at all costs.

In the animal models there is consistent evidence that, when it comes to aggressive behaviors, DHEA (the precursor to testosterone, which shares various properties with the hormone) may be more important in the brain than is actual testosterone. For example, in songbirds, levels of DHEA rise sharply in the brain following the intrusion of other birds into their territory.[95, 96] When DHEA levels are measured in the brachial vein, where blood is leaving the wing, the rise in response to intrusion isn't that significant; but when measured in the jugular vein, where blood is leaving the *brain*, the levels climb sharply compared with the rest of the circulatory system. Incidentally, injecting DHEA into songbirds makes them sing; it's a sign that these vocalizations are actually a way of marking their territory. Maybe that's why most human music is about love, too.

CUTTING OFF DEBATE

Since several studies have implicated testosterone in criminal behavior, might it be that reducing testosterone levels can help control the problem? On the heels of the attack on Jyoti Singh, lawmakers proposed a bill to legalize the castration of sex offenders, and for a time following Jyoti's death, stories of "vigilante castrations" splashed across the national news. In one case a group of women removed an attacker from a victim and immediately cut off his testicles with a meat cleaver.*

Although that may constitute viscerally satisfying retribution for some, I hope readers of this book will agree that meat cleavers are not the answer. Still, many may be surprised to hear that government-sanctioned chemical castration is alive and well in many parts of the world—including in the United States. Most of the time chemical castration is considered in the rehabilitation of pedophiles, not violent rapists, who are more likely to simply be imprisoned, but there is some overlap.

Chemical castration of convicted pedophiles is legal in several US states, including here in liberal California. Section 645 of the California penal code was signed into law on September 17, 1996, and that same day, the *San Francisco Chronicle* quoted the bill's author, Bill Hoge, a Republican assemblyman from Pasadena: "This legislation sends a clear message to child molesters—you are not welcome in California and if you commit these heinous crimes, you will be dealt with appropriately."[97] Similar laws are on the books in other states and in Europe, and numerous human rights groups have stepped up to label it "cruel and unusual punishment."

Obviously, Hoge's perspective is that these men should undergo castration as punishment. (We might say he's on the "meat cleaver"

---

*Incidentally, India still has a large class of eunuchs, called *hijras.* According to some sources, there may be as many as 1 million.

side of this debate.) Medical ethicists and human rights advocates, however, struggle with the use of chemical castration on those who commit sex crimes, and most supporters of chemical castration consider it not a punishment but a *treatment* for a biological condition that "causes" pedophilia.

Yet whether or not chemical castration is cruel and/or unusual, we do know that it can be effective, at least partially. Studies done in countries where chemical castration is also performed show that voluntary chemical castration reduced the risk of sex-crime recidivism to under 15 percent, compared to a 40 to 70 percent rate of repeat offense in non-castrated criminals.[98] There is obviously a big difference between voluntary castration and forced castration by the government, and dozens of questions swirl around the ethics of imposing punitive treatment that has such a major, permanent effect. Ethicists debate whether castration violates the criminal's liberty, or whether, in allowing him to live free from his abnormal biological urges, it actually grants him liberty. Although extreme, castration in the penal system illustrates the intersection between the horrors of misdirected virility and the potential for taming it.

Researchers have also looked into less-extreme methods of controlling negative testosterone-driven urges and behaviors in criminals. In 2005 researchers in Pennsylvania published a comprehensive study on the mental and physical effects of leuprolide therapy on a very small number of individuals. They treated a group of five incarcerated pedophiles with cognitive/behavioral therapy both before and after they were given *leuprolide acetate*, which drops testosterone levels by about 90 percent, into the "castrate" range. (This is the same drug and dose I use every day to treat prostate cancer.) To gauge the body's response, the men were then shown pictures of children (not lewd or suggestive pictures, just images like those you might see on Instagram) while researchers measured their visual reaction time as well as penile tumescence (blood flow in the penis) through a procedure known as *penile plethysmography*. Polygraph testing was also performed using questions about urges, frequency of masturbation, and other potentially relevant behaviors. The findings from these studies are about what you would expect. Testosterone levels in the blood declined by about 95 percent, the penile plethysmography units (used to measure erections) declined from baseline numbers by a little more than 50 percent, masturbation decreased, and polygraph scores came more in line with "truthfulness."[99]

In 2009 two German researchers published a systematic review in which they attempted to address all of the published studies in the field of sex-offender rehabilitation in order, at a very high level, to figure out what seems to work and what does not. In their review of more than 22,000 sex-offender cases and many controlled trials, they found that of sex offenders who were treated with testosterone-reducing hormone treatments, only 11.1 percent offended again (or at least were caught). The rate was 17.5 percent in those who did not receive any treatment. This translates to a 38 percent reduction in the likelihood of offending if the individual receives some form of hormonal treatment.[100] The results are not trivial, but neither do they point to a cure, and the authors wisely point out that hormonal treatment is so frequently accompanied by psychotherapy that the latter confounds getting an accurate readout of the former.

There is also the problem of selection bias, as in these examples the data may be skewed by the fact that men who underwent voluntary chemical castration were likely to have been more highly motivated and so more likely to show positive results. Imagine a trial of a new cholesterol medication to prevent heart attacks, but the only people who signed up to take the medication were those who already exercised and ate well in order to prevent heart attacks—you get the idea. In this case, the only way to control for such bias would be to have a test group of men who had been chemically castrated against their will, and that obviously would not pass ethical muster.

Castration laws are on the books in California, Florida, Wisconsin, Montana, Georgia, Louisiana, Oregon, and Texas, and the two constants in all state statutes are that (a) the laws apply only to repeat offenders for sexual assault of children, and (b) there is an element of choice on the part of the sex offender—in other words, there is no state-mandated castration. It is offered as just one elective component of treatment and rehabilitation.*

In Wisconsin, however, Statute 302.11 also explicitly states the following: "Inmates are entitled to mandatory release or parole after they have served two-third [*sic*] of their sentence; except the DOC [Department of Corrections] may deny the release of a serious child sex offender who refuses to participate in the pharmacological treatment using antiandrogen or its chemical equivalent." In this case, castration is not a condition of release, per se, but it is a condition of *early* release. Basically, you can be kept in prison longer by refusing to participate in the program. Is that the same thing as voluntary treatment?

In 2010, the European Commission for the Prevention of Torture and Inhuman or Degrading Treatment or Punishment (CPT) weighed in on this practice by issuing letters to Germany and the Czech Republic, the two European countries engaged in surgical castration (removal of the testicles). From the letter to Germany: "Surgical castration is a mutilating, irreversible intervention that cannot be considered as a medical necessity in the context of sexual offenders . . . Therefore, the Committee recommends that immediate steps be taken by the relevant authorities to discontinue in all German Länder [federal states] the application of surgical castration in the context of treatment of sexual offenders." The primary reason the CPT objected to the use of surgical castration was that it was, in effect, impossible

---

*In the Texas law (Gov. Code 501.0810) there is no mention of the use of reversible medical castration; the only option is surgical orchiectomy (removal of the testicles).

for a subject to give true "informed consent" to the procedure given that other incentives could influence the decision. The criminal justice system in general, and the setting of incarceration specifically, were by their very natures, they argued, settings in which a subject was unlikely ever to be free of coercion.

LIGHT IN THE DARKNESS

For now, our greatest weapon in battling criminal violence, including sexual violence, is prevention, which includes speaking out against it and building the types of laws and social supports that make it less likely to happen. In this way, an individual with excessive aggression may be more likely to seek psychological treatment or other social services as a way to fight his biological urges. When Jyoti Singh's story hit the news, it caused worldwide outrage, as it should have. In shining the spotlight on one of the darkest corners of unchecked human aggression, her death—and her parents' willingness to share her story, despite cultural taboos—has influenced the hearts, minds, and laws that will shape future generations.

As we conclude the chapter on this very ugly corner of virility, let me leave you with some hopeful statistics. The most reliable statistics available show that the rate of rape in the United States declined approximately 80 percent from the early 1970s to 2009[101] (as did most violent crime), and 58 percent between the years of 1995 and 2010 alone, and this drop occurred alongside increased public awareness and support for those reporting these crimes. We may have a long way to go, but the fact that such major changes have occurred in such a short time is further proof that we are more than the sum of our biological urges—that we can turn our backs on our evolutionary demons and successfully "nurture" our way out of what nature had in store for us.

Part III

EVOLUTION, MANIPULATION, AND ADAPTATION

Chapter Nine

POETRY AND PLACEBOS: THE ONGOING QUEST TO REKINDLE VIRILITY

Much of the west coast of Ireland is rocky and desolate. The sharp crags of the mountainous country give way to bogs and miles of limestone sheets descending out toward the cold North Atlantic. Vegetation is sparse. This area of the world has been populated for millennia by people who lived among the bogs, rocks, and fog, and who, later Christianized by the Irish, incorporated much of their original pagan mythology and practices into the religious and mystic aspects that make Irish culture unique. Driving through this landscape—even with a carful of kids, as I did a few years ago—makes for reflective travel. It comes as no surprise that this region served as inspiration for one of the great poets of the English language, William Butler Yeats.

Although born in Dublin and raised in London, W. B. Yeats is most associated with western Ireland, and especially the city of Sligo, where he lived much of his adult life and from which he drew much of his creative inspiration. In certain circles, in fact, regions of western Ireland are now known as "Yeats Country."

Prodigious from an early age, Yeats emerged on the world poetry scene in his twenties and thirties with his hallmark style of interweaving personal experiences with a rejuvenated interest in Irish mythology and the occult. He was enormously famous during his prime, and not only became a literary hero in the region but also put his stamp on the newly independent Republic of Ireland as a political entity. Quite literally, he actually designed the stamps! He received the Nobel Prize for literature in 1923, at which time the Nobel committee credited him for "inspired poetry, which in a highly artistic form gives expression to the spirit of a whole nation."

Yeats continued to write until his death in 1939, at the age of seventy-three, and he can be counted among those who enjoyed continued artistic influence as they aged. That said, contemporary scholars acknowledge that his style changed as he entered his twilight years, and the poetry and the persona of the "late Yeats" are recognizably distinct from their younger versions. In the last four to five years of his life, Yeats remade himself, and as proof of the transformation, poetry scholars and critics point to the themes of his later works, which speak to a brooding, mortality-focused worldview much in contrast to the tales of fairies and fishermen he was known for as a younger man. These later pieces also, however, channel the thoughts and observations of something of a horny old man. Here is an example from the poem "Politics," written in the poet's last years and published after his death:

How can I, that girl standing there,
My attention fix
On Roman or on Russian
Or on Spanish politics?

Scholars refer to this time as Yeats's "second puberty" (the Dublin newspaper gave him the moniker "gland old man"),[102] and the above piece seems to support this label. The question, then, is why and how did this transformation occur, and what can it tell us about the drives, biology, and psyches of aging men in general?

In the late 1920s and early 1930s, about ten years before his death, Yeats was beset by a number of illnesses, suffered the deaths of several loved ones, and felt a final, stinging rebuff from his longtime muse, Maud Gonne, a beautiful and fiery Irish revolutionary with whom he had been infatuated for decades, and to whom he proposed many times to no avail. Their relationship, one of mentorship and what they both called a "mystical marriage,"[103] had fueled his poetry for decades. Facing romantic rejection and physical frailty,* Yeats developed a new sensibility in his work, which took on a tone of melancholy. In 1928, when he was sixty-three, he wrote what is considered one of the finest poems in the English language, "Sailing to Byzantium," an expression of the poet's desire to move beyond the frailty of his aging body into a more lofty, spiritual realm. It begins with the famous line, "That is no country for old men," and goes on to explain why with lines such as "An aged man is but a paltry thing, / A tattered coat upon a stick." In the next stanza he describes his heart as "sick with desire / And fastened to a dying animal."

Some may argue that melancholy makes for good poetry, and it's true that major depression has afflicted many a great artist. Some people have suggested the depressive mentality is not only a common but a necessary accompaniment to creative genius, and it has been argued that as treatment of depression becomes more widespread, it may diminish the creative output of those affected by it. While we don't know whether

---

*Yeats eventually was married for the first time at age fifty-one, to Georgie Hyde-Lees, who was twenty-six years younger than he. The relationship with Gonne was so influential on his life and poetry that the fact of his actual wife is, literally, a footnote.

Yeats suffered from true clinical depression, the people who knew him report he went through a period of listlessness that he blamed, in part, on the loss of what he might have called his "male energies."

The complex etiology of depressive illness extends far beyond a simple relationship to a single chemical, and without doubt involves subtle derangements in the many neurotransmitters, neurosteroids, and their receptors in the brain, all of which are involved in mood regulation. That said, alterations in testosterone and its related androgens have in fact been implicated in the occurrence of depression.

Maybe Yeats wasn't suffering from clinical depression; maybe he just had low testosterone and was simply tired and melancholy. Would modern treatments have helped him feel better? I choose to highlight his life and story specifically because he lived at a time well before we could have measured his testosterone level and come up with a treatment plan; the hormone wasn't isolated until the mid-1930s, and subtly manipulating it with drugs was still some years off.

And yet, in 1934, at age sixty-eight, Yeats, looking, in his words, to "recreate" himself, sought remedy in the "rejuvenation" therapy that was popular in the day. On the recommendation of a friend, he traveled to see an Australian physician named Norman Haire, who had an office on Harley Street in London. Harley Street is to British doctors what Wall Street is to American investors or Rodeo Drive is to high-end shoppers. Yeats's objective there was treatment with the state-of-the-art therapy of the day: the Steinach procedure, or Steinach operation. Dr. Eugen Steinach's theory capitalized on the emerging science of endocrinology and the dual functions of the testicle as both (a) an endocrine organ, with the purpose of releasing testosterone into the blood, and (b) a secretory organ, reflecting its role in releasing, or "secreting," sperm. The theory was that by tying off and thereby blocking the *ductus deferens* (the sperm-outflow track), more of the testicular function could be dedicated to the production of testosterone. Today, any urologist will tell you the Steinach procedure is little more

than what we now call a vasectomy. Its effects on testosterone are min-
imal, and if it has any measurable effect at all it is usually a slight
reduction in testosterone levels, not an increase. However, this was the
thinking at the time, and the procedure was all the rage. Other patients
of Dr. Steinach's included Sigmund Freud and the eminent journalist
H. L. Mencken.

THE ELIXIR THAT LAUNCHED ENDOCRINOLOGY

The rejuvenation procedure Yeats underwent has its origins in *organotherapy*, a concept developed some fifty years prior that held at its core
the hypothesis that the testes were the source of a substance that maintained youth, vitality, potency, and virility. Testosterone itself wouldn't
be isolated until 1935—a year after Yeats's surgery. Organotherapy
proponents asserted that surgical manipulation, or even transplantation, of testicular tissue could lead to the sustenance of these positive
effects. In some cases this meant the man would receive a cross-species
testes transplant from an animal such as a goat; in other cases, the
"transplanted" element would be semen, via injection.

The originator and most vocal champion of this movement was
Charles Edward Brown-Sequard, a pioneer of this golden age of medical discovery and experimentation. His scientific life was extremely
prolific, and his observations on neuroanatomy and physiology, especially the function and structure of the spinal cord, are still taught in
medical schools to this day. He was the real deal—a giant in modern
medicine.

By the 1880s, Brown-Sequard, then in his late sixties and early seventies, was suffering symptoms similar to what Yeats would encoun-
ter some forty years later: loss of stamina, fatigue, and listlessness.
Brown-Sequard turned his research acumen to this problem and began
exploring the hypothesis that the expenditure of semen from a male

resulted in the loss of what he called a substance with "great dynamo-genic power." Curiously, he based this hypothesis on the observation of behavioral changes that occurred following sexual intercourse, and the converse observation that abstinence was associated with a preser-vation of energy. He delivered this theory to the Society de Biologie of Paris on June 1, 1889, and the paper has become known as the "Elixir of Life" lecture. In it he states:

> It is known that well-organized men, especially from twenty to thirty-five years of age, who remain absolutely free from sex-ual intercourse or any other causes of expenditure of seminal fluid, are in a state of excitement, giving them great, although abnormal, physical, and mental activity. These two series of facts contribute to show what great dynamogenic power is pos-sessed by some substance or substances which our blood owes to the testicles.[104]

He believed that the loss of semen through ejaculation resulted in a decline of male potency and vigor, and this led him to conclude that replacing the semen could renew the lost animus. In what became known as *La méthode Sequardienne*, semen and a crude testicular extract became the drug itself. Again, in his words:

> I have made use, in subcutaneous injections, of a liquid con-taining a very small quantity of water mixed with the three following parts: First, blood of the testicular veins; secondly, semen; and thirdly, juice extracted from a testicle, crushed immediately after it has been taken from a dog or a guinea pig.

His series of experiments began on himself,[105] and his observations upon starting the injections were as follows:

> My limbs, tested with a dynamometer, for a week before my trial and during the month following the first injection, showed a decided gain of strength. The average number of kilograms

moved by the flexors of the right forearm, before the first injection, was about 34½ (from 32 to 37), and after that injection
41 (from 39 to 44), the gain being from six to seven kilograms.

Working with a colleague, he subsequently injected three "old"
men (they were in their sixties) with the concoction. The subjects were
told they were receiving a "fortifying" injection and were not informed
as to the ingredients of the syringe (a gross breach of modern medical
research ethics, by the way).* Although the slightly deceptive language
was an attempt to prevent the results from being skewed by, among
other factors, the placebo effect, that's probably exactly what it produced. The men who had been given the "fortifying" treatment reported
results comparable to what Brown-Sequard himself had experienced.

For a variety of reasons, I'm skeptical that such a procedure would
have lasting results. First off, the guinea pig testicle cells would most
likely be readily rejected by the recipient's immune system and therefore limit the amount of hormone that might find its way into his circulation. Further, the semen of a stranger would similarly be rejected
by the immune system, and even though it may have contained modest
amounts of testosterone, the hormone itself would have been unlikely
to stick around in the circulation for very long—maybe a few days at
the most. That said, one cannot rule out that these men actually did
experience a transient rise in their testosterone levels and that this did
have some modest and brief effects on their mood, energy, and libido.

_____

*Although many refer to it as one of the golden ages of medical discovery, it is
important to note that the late nineteenth century did not benefit from the modern
era of informed consent of human subjects involved in research. Thus, doctors
could experiment on their patients without the need for approval of the patients,
their families, or any hospital officials. Such experimentation was common, and
although it was done with the type of well-meaning, paternalistic attitude we now
look upon with disdain ("I know what's best for you"), it would be inaccurate to
suggest that it was done with ill intent or malice. Typically, it was done with the
best interests of the patient in mind.

Most important, perhaps, was that these observations formed the foundation for the concept of hormones—circulating chemicals that arise from the sex organs—some fifty years before testosterone was isolated in the laboratory. Thus in many ways it was organotherapy that ushered in the field of endocrinology.

Regardless of what mechanisms were at work, the procedure became wildly popular, and in 1889 it was estimated that more than 12,000 physicians around the world were giving the injections, and not just for reasons of "male rejuvenation," like impotence. The Elixir of Life was eventually used, and touted, for the treatment of Parkinson's disease, diabetes, epilepsy, gangrene, paralysis, tuberculosis, hysteria, and on and on. In the United States, the Elixir of Life even showed up in popular culture. In one political cartoon, Brown-Sequard is shown injecting President Grover Cleveland, in hopes of reviving his "lifeless" free-trade policy.

Some forty years later, in the 1920s, Steinach, then the director of Vienna's Biological Institute of the Academy of Sciences, proposed that repeated injections were not necessary and that a ligation procedure could lead to sustained and possibly lifelong benefits. The procedure of tying off the *ductus deferens* rose to widespread popularity, to the point where the inventor's name was used as a verb in the modern lexicon; men who underwent the procedure were said to have been "Steinached." Like Brown-Sequard, Steinach was a legitimately accomplished physiologist. His prior work had detailed sex differentiation and advanced research on how sex organs formed during fetal and childhood development, and he was nominated for the Nobel Prize in Physiology six times during the 1920s and '30s (although he never won). Also like Brown-Sequard, Steinach gambled with his credibility late in life by espousing these procedures without the backing of the type of scientific rigor that had been the hallmark of most of his career.

Others followed with their own adaptations of Brown-Sequard's approach, including the Russian Serge Voronoff, who injected an extract made from ground monkey testicles into humans and actually performed a successful testicular transplant from monkey to human. Before moving to Paris, Voronoff had served in the court of the king of Egypt and had attended to the medical needs of the court eunuchs, who still existed in Egypt at the time. His observation of the various health problems of eunuchs* led him to conclude that the testicles held the key to general health. To test the hypothesis, he implanted the testicles of a young lamb into an old ram, which he said resulted in increased sexual activity in the ram and even a change in the thickness of its coat. From there he progressed to monkeys—presumably because of their closer genetic relationship to humans—and reported the successful transplantation of monkey testicles into humans. In California, other scientists tested testicular implants on inmates at San Quentin prison, and the experiment is said to have improved the recipients' condition and potency to the point of inspiring escape attempts. I'm not sure what those researchers were thinking, and again, this is not a case study in exemplary medical ethics.

THE CHARLATAN

As one might predict, from this movement emerged an individual who stands as one of the giants of medical quackery: John R. Brinkley. A native of Kansas, Brinkley was, before establishing his medical practice, a snake-oil salesman. Yes, he literally sold snake oil, and yes, he deserves every bit of derogatory connotation. In 1915, to improve his business prospects, he purchased a medical degree from the University of Kansas Eclectic Medical University for $500. At the time, purchasing

---

*Eunuchs are males castrated before puberty to prevent masculinization.

such diplomas rendered the purchaser eligible to practice medicine in Kansas and Arkansas.*

One of Brinkley's early jobs was that of house doctor for the Swift meatpacking company, where he was exposed to animal slaughtering and subsequently inspired by the potential of using animal tissue in medical experiments. Later, when a prominent local farmer complained to him of a decreased libido, Brinkley joked that maybe the patient would benefit from some goat glands. The farmer urged Brinkley to try it on him, and he did so. Brinkley had attended a lecture Voronoff had given in Chicago and was thus aware of the experiments happening in Europe—particularly those involving monkey glands. He saw the beginning of a lucrative business opportunity.

Brinkley promptly began the process of advertising his new techniques and acquiring testicles from the slaughterhouse—goats being, of course, much more plentiful than monkeys in Melford, Kansas (pop. 200) at that time and probably still today. As the testimonials and his advertising spread, men streamed into his clinic. Eventually, he was able to charge $750 per transplant—somewhere between $15,000 and $20,000 in today's dollars.

Ultimately, Brinkley's fame caught the personal attention of Morris Fishbein, the famous longtime editor of the *Journal of the American Medical Association*. Fishbein, speaking from a perch of substantial authority in the field, appropriately asserted that Brinkley's promises were not backed up by rigorous scientific data and that they amounted to quackery. He urged the state of Kansas to revoke Brinkley's medical license. In retaliation, Brinkley traveled to California, where he performed the procedure on the editor of the *Los Angeles Times*, who went on in that paper to extol the benefits of goat-gland transplantation.

---

*It's no longer legal to buy a medical degree through the mail—although it may be possible to do so illegally. According to the 2005 book *Degree Mills: The Billion-Dollar Industry That Has Sold Over a Million Fake Diplomas*, approximately 1 percent of US diplomas are fake, and the number is increasing.

Brinkley also adopted the new medium of talk radio to advance his agenda, producing and hosting his show *The Medical Question Box*. He would take calls from patients, administer advice, and even prescribe medications over the air (as long as they were filled by pharmacists working for the NDBPA—the National Dr. Brinkley Pharmaceutical Association), all without ever seeing the patients. His fame grew, and by the time his medical license was revoked by the Kansas Board of Medical Registration, 16,000 such procedures had been performed and $12 million in fees had been collected. In 2018, we're talking almost $300 million.

Yet, in a way, we have something to thank Brinkley for. It was his high-level deception and in-your-face charlatanism that contributed to a consensus on the need for the regulation and stricter licensing of doctors in this country, as well as the claims they, or the pharmaceutical industry, are able to make. Pope Brock's 2008 book *Charlatan: America's Most Dangerous Huckster, the Man Who Pursued Him, and the Age of Flimflam* is an entertaining read on both this topic in general and the antics of "Dr." Brinkley in particular.

POETRY—FROM A PLACEBO?

By his own account, the procedure Yeats underwent in 1934 apparently did him some good. The following four years (his last) were among the most productive of his life, and it was this period that scholars referred to as his "second puberty." It inspired such poems as "The Spur," from 1936:

> *You think it horrible that lust and rage*
> *Should dance attendance upon my old age;*
> *They were not such a plague when I was young;*
> *What else have I to spur me into song?*

Although he—and perhaps history—attributed the reinvigoration of his poetry to the reinvigoration of his body, we of course have no data on his testosterone levels. It is widely known that Yeats became promiscuous in his later years and engaged in several extramarital affairs between the time of the procedure and his death in 1939—both his wife and his current mistress were present at his deathbed. Still, while he may have been sexually rejuvenated, we can't state with any definitiveness that there was a biochemical correlate to that condition. But if it is unlikely that the Steinach procedure did anything to increase his testosterone, then what accounts for his transformation? Was it nothing more than the placebo effect?

Studies have shown time and again that sometimes placebos work, and placebo "responses" are very common in a wide variety of medical-research scenarios. The placebo effect—defined as improvement in the ailment through treatment by a substance that has no active therapeutic effect—has been observed in experiments related to treating anxiety, pain, depression, nausea, Parkinson's disease, bowel disorders, and other ailments.

Research on depression and antidepressant drugs has brought the placebo effect front and center in recent years, not only because of the ethical issues raised by giving placebos but also because many of these studies show significant changes in patients receiving placebos, among which have been lessening of depressive symptoms, reduction of suicide risk, and even changes in brain metabolism on PET scans, which measure sugar use and metabolism in different parts of the brain. In a study of depressed patients, the group given "treatment" with a placebo showed real, specific, consistent changes in the blood-flow patterns in a particular area of the brain.[106]

Scientists who study placebos point out that what we commonly refer to as the placebo effect may actually be one of three phenomena:

1. The Hawthorne effect, in which subjects act and even feel differently because they are being observed by a researcher;

2. The "ritual effect," in which the subject's experience of the therapeutic ritual (for example, of taking pills) results in reports of improving symptoms; or

3. The "attention effect," in which having the attention of doctors and the medical community—and behaving differently because of the support and nurturing that the relationship with doctors, nurses, and the medical system can (but may not always) provide—results in self-reports of improvement.[107]

It is wise to keep the placebo effect in mind when we read about studies performed in the "olden days" of medicine, when many medicines were basically inactive placebos. It is also something I, as a practicing physician, use to remind myself of the value of paying attention to my patients, including listening to, sitting with, bonding with, and touching them. One academic paper refers to this as a "contextualized healthcare response,"[108] indicating that a patient's outcome is due not to the treatment but to the therapeutic environment. Some doctors, and some health-care systems, do a better job of delivering positive environments than others.

We now have data on the anatomic and genetic underpinnings of the placebo effect. Using functional MRI scans that measure blood flow in various parts of the brain, researchers have shown that the part of the brain called the *prefrontal cortex* becomes activated during the placebo effect, and in 2012, researchers from Harvard's Program of Placebo Studies reported that a specific variation in the gene *catechol-O-methyltransferase* (COMT), which regulates the breakdown of dopamine in the prefrontal cortex, was associated with a heightened likelihood of response to a placebo in a controlled study of a medicine designed to lessen the symptoms of irritable bowel syndrome.[109] Since higher levels of the neurotransmitter dopamine generally mean more brain activity in that area, variations in the COMT gene that degrades dopamine—especially the variation that slows down that breakdown—is associated with a higher likelihood of the patient's experiencing improvements in symptoms with a placebo. The presence of that COMT variation has also been associated with other effects, including better performance on cognitive tests evaluating executive function and, unfortunately, increased sensitivity to pain.[110] It may seem like a contradiction that the gene is associated on the one hand with increased sensitivity to pain, yet on the other with improvement with a placebo. What seems to be happening is that this gene's MO is "increased response to external stimuli," whether these stimuli are real or fake, painful or pleasurable.

LESSONS FROM THE GLAND OLD DAYS

What can the experiences of the 1880s and the procedures of the 1920s and 1930s teach us about today? First and foremost, it is that the quest for male rejuvenation is nothing new. Second is that the power of an anecdote (especially from the patient of someone who, like Steinach or Brinkley, is respected in his field) can drive the adoption of medical techniques or procedures, even if the scientific evidence is not there to support the claims. In our current environment, in which social media is becoming an ever larger influence, these phenomena are important to consider. Placebo effects may lead to anecdotes, and anecdotes may be broadcast to the world as "evidence" on Twitter or Facebook through people we "know," whether or not they or their claims have any validity. When you hear a testimonial from one patient about amazing results of an unregulated product, remember: this is not science. There is no control group, so you can't know if a treatment is really creating the effects it says it can. Yeats said the Steinach procedure restored his virility, and scholars said it improved his poetry, but did it really? Or could this case be explained by variations in the dopamine in Yeats's brain as driven by a polymorphism in the COMT gene? Perhaps that gene and its increased sensitivity to pain and external stimuli opened his mind to the kinds of thoughts that inspired him to pen such moving poetry in his later years.

Finally, these stories may give us pause as we reflect on contemporary uses of testosterone as a supplement or latter-day "elixir of life." These days, male rejuvenation is happening on a large and growing scale as aging men turn to supplementation with hopes ranging from the restoration of youth to simply a bit of help with energy or depression. But is this treatment of a disease or simply "cosmetic endocrinology"? By supplementing the testosterone of the aging, are we offering a much-needed replacement of a deficit, or needlessly fighting a natural process? What we can be sure of is this: from Sligo to the Viennese clinics of the last century to the "low T" clinics popping up in medical practices and mini-malls today, men will continue to seek the rejuvenation of their youthful selves. It is probably a hardwired feature of maleness, a by-product of our most basic reproductive and survival instincts. And therein may lie a certain beauty—keeping us young at heart, yearning, and energetic. Whatever became of Yeats and his testosterone, whether a placebo effect or real, the story can make the case that, however we accomplish it, rejuvenation, and not just melancholy, makes for good poetry.

Chapter Ten

STRONGER. FASTER. HAPPIER?: SUPPLEMENTATION AND ITS DISCONTENTS

n the over-seventy-five age category, the world record for the men's half marathon is one hour, twenty-nine minutes, and twenty-six seconds. It was set in 2007, in London, Ontario, by Ed Whitlock. Whitlock was also the first person over age seventy to run a sub-three-hour marathon. How do I know that? Because, until recently, I was reminded of it every three months by one of my patients.

Marcus is eighty. Ed Whitlock died in 2017, but his record remains, and since setting the California record for the half marathon in this age group, conquering Ed Whitlock has been Marcus's playful obsession.

Marcus is on the short side, but he has a build that somehow brings to mind the term "statuesque." A former airline pilot with a chiseled chin, he slicks back his thinning sandy hair, and his nearly fat-free body is a sculpted map of musculature, showcased by the skintight workout shirt that seems to be his uniform. There is a perpetual twinkle in his eye that, while endearing, always makes it feel as if he's trying to pull a fast one. Despite the twinkle and the subtle upturn at the corners of his mouth, he only rarely cracks a full smile, yet his appearance and energy suggest that he's having the time of his life. I'm not sure if that's actually the case or if it's only what he *wants* me to think, but his visits are fun for both of us, and we have a little ritual every time he comes in. First, he gives me an affectionate "one-two" of soft punches to my gut and I feign doubling over. Then he looks me in the eye, his eyebrows rise, and he mouths the words "one," "twenty-nine," and "twenty-six."

I took this lofty goal of his seriously at first, and wouldn't have dreamed of making fun—and then I realized he likes sarcasm as much as I do. These days I look him straight back and say, "Yeah, right!"

We both laugh. This jovial pushback is my contribution to our game. He knows I support him and would love to be there when he breaks the record—but we both know he probably never will, as every year he gets a little older and his dream a little less likely.

In a practice that sees all too much suffering from cancer, thriving patients like Marcus are a breath of fresh air. He is fortunate in that his cancer is not active, cured a few years back by a combination of hormonal therapy and radiation. He comes to me now because he wants me to monitor his PSA value while he takes supplemental testosterone.

Marcus has developed a system that works for him: apply a topical testosterone gel under the arm Thursday through Saturday, then take the rest of the week off. While using this formula, he's been able to complete several half marathons, which he trains for by running three or four miles every day and following that up with an hour-long workout in the gym. He likes to tell me that his younger wife has trouble keeping up.

He's been on this regimen for a few years, ever since his testosterone failed to normalize after the androgen-deprivation therapy he underwent as cancer treatment. Many patients recover their testosterone naturally after a year or so, but some, especially the older ones, don't. Truthfully, Marcus doesn't need to see me as often as he does, but I think he gets some reassurance from our sessions, so I let him come whenever he wants. Besides, his enthusiasm for life is infectious.

SUPPLEMENTATION AND SECOND ACTS

Ask random people what they think of when they hear "testosterone supplementation" and you're likely to get an answer that includes the words "erections," "sex," and/or "muscles." We'll get to all three eventually, but let's tackle erections first, in part because their relationship to testosterone is probably the most misunderstood. Many people confuse supplemental testosterone with drugs like Viagra, believing they are or do the same things. They aren't and they don't.

As you know by now, testosterone is a naturally occurring hormone made in the testicles and adrenal glands that circulates through the bloodstream and binds to receptors in a variety of spots in the body, with all kinds of fascinating and complicated effects. Erectile-dysfunction treatments like Viagra, Cialis, and Levitra, on the other hand, are *not* naturally occurring substances, and in fact have nothing to do with hormones at all. Their main action is simply to promote blood flow to the penis. They are a type of *vasodilator* (a drug that relaxes the walls of blood vessels), which targets a specific type of vessel mostly found in the penis.

The clinical effects of supplemental testosterone are far less straightforward. A large study recently published in the *New England Journal of Medicine* revealed the results of a placebo-controlled trial of supplemental testosterone administered to older men who had either confirmed low testosterone or symptoms of low testosterone. The treatment was the same formulation as the one Marcus takes, but was given daily. The results showed that supplementation for one year boosted libido and mood but, contrary to what Brown-Sequard found with his dynamometer, the higher level of testosterone didn't actually contribute much to physical strength or stamina as measured by a test of walking ability and a questionnaire on fatigue used to measure the vague but important concept of "vitality."* Also, while libido improved briskly at the beginning of the trial, by one year the effect had begun to wear off. Libido scores improved by 60 percent after three months, but by twelve months they were just 25 percent above the baseline value.[111]

Interestingly, a combined study by the NYU and Mount Sinai medical schools a couple of years ago showed that beta endorphins—our bodies' natural morphine molecules—almost doubled during anabolic-steroid use. Testosterone, of course, is an anabolic steroid, which might explain why people feel so good *emotionally* while taking testosterone supplements.[112] It might also explain why the use of testosterone and other steroids by athletes is so often compared to addiction.

Is it possible that there's a disconnect between how a high testosterone level makes Marcus *feel* and the effect it actually has on his physical condition? If supplemental testosterone for the aging male needed a poster child, Marcus would be it; but I wonder if the supplement is truly driving Marcus's energy and enthusiasm or if that's just who he is. I also wonder whether this testosterone supplementation is truly *restoring* his testosterone levels to where they "should" be or if taking him to a higher level than is "normal" for his age goes beyond restoration. After all, most men of eighty years have a testosterone level that is about half of Marcus's. *Andropause*—the gradual decline in testosterone in aging men—is increasingly being viewed as a medical condition meriting treatment. But I must ask: Is this a condition we *should* correct? If so, what is the cost of that correction—not in terms of dollars but in terms of biological and/or psychological side effects? For instance, some

---

*"Vitality" in the context of quality-of-life research is a composite score that comes from a questionnaire called the Functional Assessment of Chronic Illness Therapy (FACIT). Statements such as, "I have trouble finishing things because I am too tired," or, "I have to limit my social activity because I am tired," are scored on a scale of 0 (disagree) to 4 (strongly agree).

studies have suggested that testosterone supplementation increases the risk of a heart attack and strokes, while others say it decreases such risks—this question is the subject of ongoing research.[113]

Marcus is not alone in his desire for performance and passion, and for many men what we might call "cosmetic" testosterone supplementation is the key to finding their "fountain of youth." History is full of comic and even tragic examples of aging men striving to regain their lost mojo, and yet our eyes still light up at the promise of feeling like we did when we were in our prime. But should every man, young or old, have access to testosterone supplements? It's impossible to draw a solid line between what constitutes legitimate medical use and what one might call cosmetic use. Many men do need testosterone to treat specific medical problems, and the FDA can't effectively regulate how doctors prescribe it, to whom, and for what reasons.

THE ROMANCE OF REVIVAL AND THE REALITY OF RISK

Marcus is clearly enjoying himself, but there's no saying whether what works for him will work for any other person. And nobody truly knows whether supplementation leads to a net harm or a net benefit in an aging population; as with many medical treatments, it may benefit some and harm others, or simply have no effect. Any doctor who tells you these sorts of decisions are easy or that the data is clear is either a fool or trying to sell you something. And there are a lot of people out there trying to sell you something.

In the midst of controversy over treatment options, prescriptions of testosterone filled by US pharmacies increased *tenfold* between 2000 and 2011.[114] The number is shocking, and it may still grossly underestimate how much testosterone Americans are taking, given that many patients receive their drugs from online pharmacies in other countries (often Canada), and even more take unregulated supplements containing testosterone-like chemicals. One recent study discovered that over half of those who use these supplements buy them online. An anonymous survey of users at a fitness center revealed that many prefer more and more potent steroids, in some cases using *trenbolone*, the powerful synthetic testosterone used to stimulate muscle growth in beef cattle.* Encased in a veneer of legitimacy lent to it by people who need it for medical reasons, testosterone supplementation remains a keystone of "wellness" programs touted by primary care doctors and men's health specialists.

Many men receiving even prescribed testosterone supplementation may not have levels below normal limits, or may be just on the low end of the normal range. I spoke with one gentleman in his seventies who told me he was given testosterone as part of a "weight-loss boot camp" program developed as a side business by his primary care doctor. He wasn't informed of any potential downsides. Two years into the treatment, he began to have pain in his shoulder, and a medical exam revealed prostate cancer that had spread to his bones—perhaps not caused but surely fueled by this supplementation. His cancer is incurable, but fortunately he is doing well on hormonal therapy.

I can't speak for what he was like in his childhood, or even throughout much of his adulthood, but in his golden years, Marcus strikes me as a guy who's driven by testosterone. We've joked about this a bit: "My wife would agree with you on that!" he's said. One interesting detail is that Marcus doesn't mention his wife much in conversation. Instead, the focus of our discussions is frequently his impressive sexual performance. Sometimes, listening to him talk about his "romantic life," I think I should refer him to the advertising agency for those low-testosterone commercials you hear on the radio

---

*Beef cattle are stimulated late in life to boost muscle growth. As a result, the runoff from feedlots frequently contains excreted hormones. A study in Nebraska found that the reproductive cycles of fish in nearby rivers were also affected, with the result that they became sterile. Another argument for vegetarianism.

or see on TV. He would actually be a good spokesman—all eighty years of him. The language of those ads is interesting, as they promise improvements for a man's "romantic life" or "intimacy," but given the action of these medications, these terms are often only a euphemism for sex. When I hear these ads, I sometimes gently suggest to my car radio that "intimacy" and "romantic life" are not the same as sex, and in fact they have different hormonal drivers—oxytocin for one and testosterone for the other. I don't think those details have been totally worked out by the scientists yet, and I can't pretend to be an expert on the subject, but that doesn't stop me from talking back to commercials during my commute.

After a year or so of testosterone supplementation, Marcus and I are hit with a dilemma: his PSA value is rising. Have we reawakened the cancer? It's too early to know, but I have a duty to suggest to Marcus that there could be a relationship between the testosterone supplementation and the rising PSA. It's not a rapid rise, and the absolute level is pretty low, but still.

"And what if I just stay on it?" he asks. It's a reasonable question, but I can see the wheels turning as he asks it.

"Maybe nothing for a while," I say. "On the other hand, this cancer is driven by testosterone and it is possible that we are waking up some sleeping cancer cells." At this point I'm thinking of the man who took testosterone in his weight-loss boot camp and only discovered his cancer after it invaded his bones.

Unlike Aaron from Chapter 1, who underwent surgery to remove his prostate and should therefore have had a zero PSA level, men like Marcus, who were treated with only radiation, may still produce some PSA in the prostate, which is still in there, only shrunken. Given that, we can't be certain that the rise in Marcus's PSA is actually due to cancer, and it's on that basis that we decide it is reasonable for him to continue the testosterone gel as long as we monitor his PSA very closely. He clearly benefits from supplementation, and it's obvious his quality of life while taking it is pretty good. Plus, he has Ed Whitlock's record to chase.

As you might have sensed, I'm a bit ambivalent about Marcus's use of testosterone, although I'm confident we'll be able to control his PSA, even if the levels do rise higher. He's otherwise healthy, and the current treatment is letting him get the most out of life. Fortunately, Marcus is not just in good physical shape, but he is mentally healthy, and the only questionable behavior we could possibly blame on the testosterone use is his cheerful obsession with breaking the half marathon world record for his age group.

Putting his individual case aside, however, I do worry about the emergence of a society that seems to be, in ever greater amounts, pouring the hormone indiscriminately into aging men, not to mention young men who undertake testosterone supplementation on their own. Used by the wrong person for the wrong reasons in the wrong place and at the wrong time, supplementation could do more damage than good.

BIRDS AND BELL CURVES

In my clinical world, we obsess about the role of testosterone on the growth of cancer. We target the levels of testosterone in the blood and in tumors, and we target testosterone's "on" switch—the androgen receptor—with drugs that are being synthesized to be ever more potent. One persistent theme that has emerged in both the literature and medical practice of hormone treatment, however, is that more is not necessarily better; testosterone levels follow their own sort of biological bell curve. If you take a prostate cancer cell and grow it artificially in a lab, you will see an interesting phenomenon: little testosterone leads to little growth, and more testosterone leads to more growth, but *even more* testosterone leads to *diminished* growth. Yet I have colleagues who are doing clinical trials of high-dose testosterone in men with advanced prostate cancer, and in some patients it works! Before we get to how and why it works, let's compare this effect to a similar one observed elsewhere in the animal kingdom.

As a professional doctor and an amateur observer of natural science, I am struck by this bell-curve idea of little, not enough, more, just right, more yet, too much, and I was excited to read an academic paper suggesting a bell curve related to testosterone might apply not just to humans but to other complex organisms. Again, we look to songbirds, and in this case the mating patterns of the *Junco hyemalis* (dark-eyed junco), a small sparrow found throughout the forests and backyards of North America. Researchers in Virginia captured several of these birds and injected them with varying concentrations of testosterone, and then followed their breeding behavior.

In a nutshell, the results of the experiment were that little testosterone led to very little mating, while more testosterone led to more mating (both within pair and extra-pair*). However, when they pushed the testosterone dose higher—beyond the "sweet spot"—mating behavior died off.[115] Clearly, the effect of testosterone is not linear. Why?

We think there are molecular damping effects in certain cells that shut down the response to testosterone when the levels rise too high.†
While this mechanism may exist within individual cells, however, it is a big leap to argue that this damping effect is not only present but also strong enough to account for the behavior of complex multicellular organisms with a frontal lobe, like humans.

From the perspective of evolutionary anthropology, one of two things may be happening: either the energy from the high levels of testosterone is being applied to male-to-male competition instead of mating, *or* the high testosterone is "turning off" the females, who then

---

*"Extra-pair" mating is ornithology-speak for sleeping around.
†In the case of prostate cancer cells, that would slow down, or even stop, the growth of the cancer.

decline to mate with the pumped-up males. In human terms, it's as if all the high-testosterone guys get into a bar fight and the girl finds a non-fighting man to leave with—probably one with just enough (but not too much) testosterone.

THE SELFISH MOLECULE?

You might say those fighting bar bros had become so focused on appearing dominant that they lost sight of the reason they were driven to do so in the first place. Indeed, selfishness has been reported in men with high levels of testosterone. For all his friendliness, one impression I often get from Marcus is that "it's all about him," and while I'm glad he likes how he feels, I can't help but wonder if this, too, could be a result of the supplementation.

As we've seen throughout this book, testosterone is not just about sex and muscles and competition. It's about empathy, identity, the moral choices we make, and so much more. It is even about politics.

The election season of 2016 unfortunately widened the divide between Americans on different sides of the political spectrum, and as I wrote this book during the thickest months of Trump versus Clinton, I couldn't help but see how testosterone and its effects were at play in the campaigns. This is, of course, the election in which penis size was actually debated by candidates on national television. But rather than devote any more ink to that angle of the relationship between testosterone and politics, let me consider how testosterone might be playing another role, one perceived not as aggression but as selfishness: in criticisms of the conservative right, the left-leaning rhetoric frequently attempts to cast leaders and candidates on the other side as selfish in one form or another. How many ads do we see where the candidate on the right is criticized for selfishly "protecting rich friends on Wall Street" or pandering to groups whose top priority is making sure they "get theirs"? Conversely, the right criticizes those on the left as "bleed-ing hearts" who are, for instance, too generous with their "handouts" and too accepting of immigrants hoping to make better lives for them-selves in our country. Perhaps the dividing line is that one side values the ability to make firm decisions based on a framework of black-and-white "moral clarity," while the other operates on the politics of inclu-sion, as driven by empathy. The difference seems like something we could measure—CAG repeats and 2D:4D ratios in Republicans and Democrats—but so far I haven't heard of anyone attempting to do so.

And of course it isn't really that simple. Some of the most gener-ous people I know call themselves political conservatives, and I have known plenty of liberal-leaning acquaintances to be, well, selfish and ego-driven. Politics turns these traits into caricature, but in reality, all of us have all of these traits, and we cycle through them depending on the circumstance, our history, and dozens of other factors. And yes, hormones may well be one of them.

In an academic paper that began with the question, "How do humans decide when to be selfish or selfless?" economist Paul Zak, from Claremont University in Southern California, conducted an experiment that showed testosterone makes us less generous—about 27 percent less generous, to be precise. His team demonstrated this with a double-blind placebo-controlled study of subjects who were administered testosterone before engaging in a psychological test of generosity and selfishness known as "the ultimatum game."

In this game there are two players, called "Decision Maker 1" and "Decision Maker 2" (DM-1 and DM-2). In the exercise they are physically separated from one another and can't directly communicate. DM-1 is given $10 and is asked to share an amount of his choice with DM-2. If he offers exactly half to DM-2, then they both get $5. If he offers some other amount and the offer is rejected by DM-2, then they both get $0. The test evaluates the minimum that DM-2 would accept getting and the maximum that DM-1 would give.

The (all male) test subjects played the game two times, six to twelve weeks apart, once after getting a dose of testosterone gel the night before and once after getting a shot of placebo. After treatment with testosterone, the average total testosterone level rose by about 60 percent, and levels of DHT—the more potent derivative of testosterone—went up by 128 percent. Levels following placebo injections stayed flat.

Consistent with what you might think, the proposals coming from DM-1 were lower after the testosterone shot, showing they were less generous. Also, the rates of rejection by DM-2 were higher, showing they were greedier. The main outcome of the study—the proposal-to-rejection ratio—was reduced by 27 percent. To quote the paper directly: "Participants on AndroGel were more than twice as likely to have exhibited negative generosity compared to themselves on placebo."

Negative generosity? Isn't that the Orwellian newspeak for "selfishness"? As it turns out, in the printed version of the journal, the "running title" of the paper, printed in the heading, is "Testosterone and Selfishness."

Sometimes what's most interesting about studies like this is not what happens to the average of all subjects but what happens at the margins—at the ends of the bell curve. In this study, when Zak and colleagues looked at the highest versus lowest levels of DHT in the blood, they found that the subjects with the lowest one-tenth of DHT levels were a whopping 560 percent more generous than those in the top one-tenth, at the other extreme.[116] The patients I treat with androgen-deprivation therapy live in that bottom tenth.* I wonder where Marcus would fall? It is hard to say.

---

*For a couple of decades now, prostate cancer research has been known as a field in which there is a large amount of philanthropic support. I am the beneficiary many times over of the largesse of grateful patients and their families, who give generous amounts of money either to my institution or to groups that support our research, such as the Prostate Cancer Foundation. Gifts from donors have paid for clinical trials, supported laboratory experiments, and helped us train and encourage the next generation of young doctors to devote their careers to research.

My main criticism of Zak's study and similar psychological research is that the subjects were twenty-one-year-old college students. This is often the case in these studies, as undergraduate students are relatively captive on college campuses and like to volunteer for research studies because they often get a little cash in return. When I was twenty-one and working in the lab at the University of Michigan, I once got sixty bucks for letting someone put electrodes on my eyeballs and then stick my head inside a sphere with flashing lights for an hour—quite an inauspicious beginning to my life in medical research. In contemplating what Zak's experiment tells us about testosterone in general, we must recognize that the twenty-one-year-old brain infused with supplemental testosterone may act a bit differently from the fifty-year-old brain similarly treated, not to mention the eighty-year-old brain of Marcus. I'm not aware of research similar to Zak's that studies the same questions in older men, and so for now I can only wonder how the supplemental testosterone may be affecting feelings and behaviors related to selfishness and generosity in the men I treat.

HOW LONG DO YOU WANT TO LIVE, ANYWAY?

After a couple of years of fun visits, half marathons, and ribbing over world records, Marcus began questioning his use of testosterone supplementation. His PSA blipped up a couple of notches—nothing serious, but it put some fear into him. I'd always had some questions about what the best course of action was in his case, and now Marcus, too,

---

Cynics have suggested to me that the reason philanthropy in this field is so bountiful is that it's "all about old men worrying about getting erections." But fears of mortality, pain, and suffering drive much of this bigheartedness, as does gratitude. Yet, perhaps we can't completely discount the potential effect that testosterone manipulation can have on the generosity of our donor patients and the empathy they may feel for other current and future patients with this disease. Whatever is behind it, we remain thankful for their support.

said he wanted to think about it. I noted our collective ambivalence in his medical record and told him to come back in three months, when we would discuss it further.

After that, Marcus simply disappeared. No visits and not a word from anyone.

In my work it's always a little worrisome when someone stops coming in for periodic visits, as it usually means one of three things: he's fired me; he doesn't need me, or—the worst—he's gone. Considering Marcus's age, I knew the last of these was a distinct possibility, and yet he was in such good shape that it wasn't my first guess. Besides, I figured we probably would have received a call from his wife or at least his primary care doctor.

I have hundreds of patients, and it's sometimes hard to track them between visits, yet certain people stick out, and Marcus was one of them. I had already penned some of this narrative by the time that last visit occurred, and a couple of times I thought of calling him just to check in. Perhaps he was traveling, or had some family issue or even another, unrelated medical problem. Perhaps he was chasing Ed Whitlock on some country road somewhere.

I wasn't upset. I let it go.

Oncologist, bioethicist, and architect of the Affordable Care Act Ezekiel Emanuel started a controversy in 2014 with a piece he published in the *Atlantic* titled, "Why I Hope to Die at 75."[117] He is in his fifties now.

He spelled out his argument step by step, pitching his position contrary to what he called the ethos of the "American immortal," meaning the belief of many aging Americans that they should, and are in fact entitled to, live as long as is physically possible. He proposes that, while death is certainly a loss, living too long also constitutes a form of loss; for him, merely surviving in a faltering, failing, and increasingly debilitated body is a greater indignity than death. Yeats was onto this idea eighty years ago when he called an aging man "a tattered coat upon a stick."

As Emanuel put it, "I reject this aspiration. I think this manic desperation to endlessly extend life is misguided and potentially destructive." He also noted that the application of medical and societal resources to those at the end of life necessarily diverts them from the young, who may have greater potential and opportunity to do more good for society as a result of medical intervention. He contends that the American immortal is destined for ever-longer life expectancy but that the years gained will be lived with a high likelihood of disability or some other functional limitation. In short, although the dream is of a "straight edge" existence—living life fully, happily, and healthily right up until a sharp decline and death—for the most part what actually happens is a gradual, progressive, and sometimes painful descent into chronic suffering before death. And for what?

The article sparked numerous debates, and of course that was the point. Whether his stance has any chance of gaining traction in the populace as a whole, however, is doubtful. The concept of the American immortal is still viewed as an ideal, and the biotechnology and pharmaceutical industries—along with businesses and individuals including Google and Amazon's Jeff Bezos—continue to invest billions of dollars in the idea that we can "cure death" and push life expectancy out to 120 years or more. The question is whether such efforts move us toward extending our quality of life or just our quantity of years. And how does this question apply to my patients? My goal is to give them both longer lives and lives of better quality. Does this make them American immortals? I happen to know Marcus is one—and proudly so.

RETURN OF THE REVENANT

One day, amidst my mulling and worrying and remembering and then forgetting to call him, Marcus showed up on my schedule. It had been more than a year.

He was doing fine and looked great, just a year older. No, he hadn't run any half marathons; he had at last given up that goal. He'd also given up testosterone supplementation, obviously, and his testosterone was down, not in the "castrate range" where it had been during treatment, but a bit below normal. His PSA was also down, and there wasn't anything to suggest that there was active cancer.

Still, there was a significant change. For the first time, he asked about my kids, their ages, and what they were up to. And then Marcus opened up more than he ever had. From what I could gather, he'd mostly been busy during the last year being a dad and a grandfather, and his main concern was his forty-six-year-old daughter, who had recently been through a difficult divorce and had been diagnosed with diabetes, all while raising two kids. His time had been taken up with getting her and the grandkids settled into a new home, and helping her navigate both medical and legal entanglements related to her sudden life change. He was the pick-up and drop-off grandpa now, too, shuttling his grandkids to school and sports practices. In a minivan, no doubt. For the first time in our talks, he referred to himself in the third person—as "Papa the Taxi Driver."

It was funny; before that visit I couldn't recall his even mentioning that he *had* a daughter. Suddenly, it *wasn't* all about him. I may have been biased by what I'd read about empathy, selfishness, and testosterone while researching this book, and yet I swear that Marcus displayed more empathy during this one visit than he had during all the years he was on the testosterone. I can't prove it, but it struck me.

In his piece for the *Atlantic*, Ezekiel Emanuel saves himself from a tone of dour pessimism by suggesting that old age is something to be celebrated for what he maintains is its true function: nurturing of the young. Or, as Emanuel prefers, "mentoring." To paraphrase, he says that by reaching out to the young among us, to those who need the guidance and wisdom of the aged, we rescue the purpose of aging. Although having low testosterone might slow Marcus down, he seemed happy, and I couldn't help but feel that this Marcus was exactly as he should be: a gentle, thoughtful, and caring father and grandfather. He'd gone from being the guy chasing an unattainable world record to just being Papa, and all seemed right with the world. Life is full of contra-dictions. Coming off the medicine he'd used to help him live as if he were younger may now, in some way, have helped him live *better*.

Marcus, my friend, you've found your place. You've been given the opportunity Alzheimer's disease stole from Warren, and cancer stole from James and so many others: to nurture the young, impart your wisdom, and share your love with those who cherish your presence.

Let someone else chase Ed Whitlock.

Chapter Eleven

BEAUTY VERSUS THE BEAST: BODYBUILDING, STEROIDS, AND THE AESTHETICS OF MANHOOD

You may not be aware (or care), but there is something of a civil war raging in the professional bodybuilding community. I know there are lots of issues in the world right now and it can be hard to keep up with the Mr. Olympia competition and its ramifications on a day-to-day basis, but I am here to tell you that this group has something to teach all of us.

The war in bodybuilding is being fought between those who believe that it is an *aesthetic* endeavor, an art form, versus those who believe that it is about chasing limits—pushing the body into greater sizes and proportions, beauty and art be damned. And this is important. Drawing the battle lines can help us understand a bit about beauty, sports, and, of course, the role of testosterone in the world. But at its core, this is a story of body image, what it does for us, and what it can cost us. And how virility and testosterone stir the pot.

PERFECTION REFLECTIONS

To set the stage, I'll start with a personal story and a bit of a digression. I recently returned to my high school in Appleton, Wisconsin, for an alumni function, and when I was given the opportunity to invite someone to the event, my former football coach was the first person who sprang to mind. Coach Engen embodied just about everything good that can be said to come of involvement in sports: he taught me about teamwork, setting goals, mental toughness, you name it, and he has since been inducted into the state football hall of fame and is revered in the community. At the beginning of my senior season, he passed out printed cards to us that read, "I may be only one, but I am one. I cannot do all things, but I can do some things, and that which I can do I will do." Very Zen, in a humble, Midwestern kind of way. I had the card taped to the inside of my helmet the entire season, and I don't know how many times I read those words, but it must have been a lot because thirty years later I didn't even hesitate as I typed them out here.

Although the strategic placement of that card in my helmet did not protect me from the concussion I got in my final regular-season game, it has likely helped me through a few challenging moments, including long nights of study, difficult challenges with patients, and in general keeping some perspective about my place in the world—understanding both my potential and my limitations.

I always wanted to impress Coach, and thirty years later he still intimidated the hell out of me. At the reunion, I semi-lied, telling him, "I still lift."

And he said, "So do I." (I knew he was telling the truth; he's in his midseventies and looks great.)

Back in high school, "lifting" was about many things. It was about strength, of course, and it was about preparing for competition in sports, but it was also about goal-setting. We even had a tribe of sorts—the "200 Club," into which you were inducted if you could bench press two hundred pounds. In our high school football program, attaining 200 Club membership status came with a T-shirt to be worn proudly, and your name was added to a board on the weight-room wall. In junior high school I revered those shirts, and four years later I had my own. I look back fondly on this early experience of setting and attaining a goal, a lesson that is valuable for anybody in any endeavor.

But although we engaged in these activities for the benefit of our bodies, we weren't really thinking about our *health* too much back then. We were adolescent boys; cardiovascular fitness and its effect on aging were not high on our list of concerns. We wanted to be strong for sports, but we also, of course, cared about our appearance. Although we wouldn't have admitted it at the time, for some of us it was even about *beauty*.

Yes, beauty.

At this point, reader, I hear you asking, "And what about vanity, Dr. Ryan?" OK, I'll concede that, too, but I will in turn challenge you to try to separate vanity from the quest for beauty, especially in the average teenager or young adult. At that age, the two concepts are pretty much woven together into one garment; virility, body image, competition, mate attraction—everything intertwined with everything else.

For the most part, the aim of staying fit is self-improvement, but unfortunately there's also a dark side: the land mines of low self-esteem, body-shaming, and eating disorders. In truth, it can be hard to avoid the urge to constantly judge oneself and others in a social context, and it can be even harder in the physical context of the gym. Our weight room, like so many others, featured one whole wall of floor-to-ceiling mirrors, making it nearly impossible to concentrate entirely on health and fitness without also acknowledging some looming aesthetic ideal.

But what "should" our bodies look like, anyway, and why do we care so much about it? Why do some people starve themselves or push their bodies beyond healthy limits in order to look a certain way? Is the idea of a "perfect" body just the product of our shallow modern society, or does it have some evolutionary significance? And is it even possible to answer these questions considering that, to use an old phrase, beauty is in the eye of the beholder, differing not just among individuals but across cultures and eras? Consider that whereas human history once linked the muscular, virile male body to manual labor—and even celebrated obesity as a sign of affluence and freedom from that labor—the trend for the past few generations has turned this concept on its head. Now, buff, muscular, and lean bodies are associated with success and even affluence while obesity is largely looked down upon, both due to the link to poor health and out of a pejorative connection to laziness and even poverty. Media outlets are constantly telling us fit people are superior, and discrimination against obese people remains rampant, in part because this group is not well protected by anti-discrimination laws. Further, this very discrimination has been shown to worsen the mental and physical health of those subjected to it.[118] At the end of the day, it's important to remember that aesthetic tastes are in part shaped by society, and there is no one true ideal.

In an effort to quell concerns about body inadequacies, a British medical website called *The Online Doctor* sponsored a study called "Perceptions of Perfection" in which the authors illuminated the wide global variation in what is considered an "ideal" male or female body.[119] The test subjects were a group of freelance graphic designers (male and female) from nineteen countries around the world. In the experiment on the male ideal, each designer was given the same photograph of a thirtysomething bearded white man with a not-altogether-fit body and told to Photoshop the image to produce the most attractive man they could. The results were quite fascinating. Some designers hardly touched the image, while others made creative changes, such as the dark highlights around the eyes and purple patterned shorts added by the designer from Bangladesh. In Russia the ideal male became more muscular and had wavy, windblown blond hair. In Egypt and Nigeria his skin was darkened, as you might have expected, but in other countries where you might also have expected that change in skin tone (e.g., the Philippines and Bangladesh), it wasn't. In many countries the man was made slightly more muscular, but the most notable changes to physique happened in the United States. The US redesign resulted in washboard abs and vein-popping biceps, and to my eyes, was the most dramatically manipulated. Ironically, considering our country's relative rate of obesity, the US ideal also appeared to have the least body fat. Overall, the physique of the US ideal seemed the least likely to be widely attainable.

I, for one, am not afraid to admit that when I see an image of a shirtless male with what I would deem a perfect body, I feel a twinge of longing. It's neither jealousy nor sexual attraction but rather a sense that there *is* an ideal male form out there and that somehow, because I don't have it, I am flawed. It is an *aesthetic* longing. And yet there is truth behind the expression "too much of a good thing." If you push certain attributes to an extreme, you're in danger of pushing them right off the edge. Think about the person who is *too* tan or *too* thin or *too* muscle-bound. Both evolution and contemporary society have encouraged us to value a man's strength, and yet many people, and perhaps especially and most ironically women, are repulsed by the modern bodybuilder—bulging out all over and ripped to the extreme. Again, it seems we have another bell curve: on one end is the male body with hardly any muscle, and at the other end is the beefed-up weight lifter. In the middle, we have a wide range of bodies that the largest number of people would find appealing. It's a kind of Goldilocks moment: these bodies are not too weak, not too strong, but "just right." Evolutionarily, there's also a connection between looking good and being healthy: within a certain range, low body fat is good and strength is good, and therefore looking strong and healthy is considered "beautiful" to both men and women for a reason—it probably *is* strong and healthy.

Pondering the concept of beauty is amongst the oldest of humanity's philosophical pastimes. In women, beauty has traditionally been linked to femininity (although that is changing), and in men it has been, and probably always will be to some degree, inextricably linked to virility. Continuing with the calculus-like graphing of these concepts, let me suggest that while there may be a bell-shaped curve to our *perceptions* of what makes a body beautiful, in terms of the body's *physical potential*, the curve is parabolic. On the far end of the curve, bodybuilders strive toward the asymptote—trying to get as close to the absolute limit of the human body's potential as it is possible to be. As they do so, the curve flattens out, which means they're putting in more and more work but getting less and less gain, expending greater effort for smaller improvements. And there's the challenge: hard work can only get you so far. A person who wants to move ever closer to the edge may look for ways to augment his natural gifts and abilities. One popular form of enhancement is—you guessed it—testosterone supplementation.

ARISTOTLE SCHWARZENEGGER

In 2015, after the completion of the Arnold Classic, now considered one of the world's premier bodybuilding events, the former Mr. Olympia, movie star, governor, and host of *The Apprentice* (is there a more eclectic resume?), Arnold Schwarzenegger, took the microphone and addressed the crowd. He expressed his dismay at the lack of appreciation for form, beauty, and aesthetics in the current bodybuilding world. Referring to many of the competitors of today, he said (and please imagine this in his famous accent) that they "do not look as pleasing. It is not acceptable. Too big, too enormous, no talent behind it."

He specifically cited the competitors' abdominal muscles, pointing out that it was now common to have the "stomach sticking out." "It doesn't look right anymore," he added. "Proportion is gone."

And with that, he demanded that the judges of the bodybuilding community reevaluate the criteria by which contestants were judged. "We don't want to see the biggest guy; we want to see the most beautiful," he said.

So what are the criteria for beauty in the world of bodybuilding? Among the constants are symmetry and clear definition of muscles, with striations visible. The term "dry" is also used a lot, referring to hydration status that presents itself as lack of general bloat. The goal is for the muscles to show through the skin, without being obscured by water and body fat. Look up pictures of Arnold in his prime (circa 1976), after he had won *seven* Mr. Olympia competitions. What you're seeing is one of the best examples of a bodybuilder focused on aesthetics.

Even though he's a household name, face, and accent today, the influence of Arnold Schwarzenegger on the world of bodybuilding cannot be overstated, which is why his admonishment of the direction the sport had taken sent ripples of discussion and debate through the entire bodybuilding community. Before retiring to pursue his acting career, he reigned over the sport during its heyday in the 1970s, and his influence crystallized and epitomized the centrality of aesthetics in the sport and formed the foundation for judging for the next twenty years or more. The movie *Pumping Iron*, released in 1976, is a touchstone for many in the sport, and it is no accident that the opening of the movie reveals a twenty-eight-year-old Schwarzenegger in a ballet studio; for him and his peers, it wasn't about strength alone, but about beauty.

If asked to define beauty, the average American person would probably offer some version based on the definition in Aristotle's *Poetics* (even if they've never read it). The ancient Greek philosopher wrote, in about BCE 335, that "to be beautiful, a living creature, and every whole made up of parts, must . . . present a certain order in its arrangement of parts." In *Metaphysics,* he expanded on the defining features of beauty: "The chief forms of beauty are order and symmetry and definiteness, which the mathematical sciences demonstrate in a special degree." This is what the aesthetic bodybuilders and the judges in the sport espouse: order, symmetry, balance, and mathematical proportion.

On the other side of the debate are those who think the goal of the bodybuilder should be to push his physical form to its absolute limits. One of the icons of this set was Rich Piana, who died recently, at age forty-five, of complications after a heart attack. After he retired from the pro circuit he maintained a massive social media following. His goal was simple: be huge, whatever it takes. And he took a lot. He began using steroids in his teen years, for which he made no apologies. For Piana and his followers, proportion, symmetry, hard work, and natural ability . . . everything is secondary to size.

These extreme bodybuilders are pushing the limits of the human body beyond what evolution would have us be, and in order to do that they use a toolkit of pharmaceuticals, one of the most popular being industrial-grade, high-dose testosterone. Does it make their muscles giant? You bet. Does it also cause a variety of health issues? Absolutely. Their joints and hearts struggle to deal with the extra mass, and Rich Piana is far from the only extreme bodybuilder to die young. Published autopsy results of steroid-abusing bodybuilders typically show fibrosis in the heart—scar tissue between the muscle cells that leads to poor heart-rhythm conduction, poor heart-pump function, or both. For those who live to see old age, I can't imagine their bodies are healthy or comfortable. Some of the substances bodybuilders take, like human growth hormone (HGH), can even stimulate the growth of cancers.

To get a glimpse into this world, I checked in with Doug, a regular guy who dipped his toe in the pro-bodybuilding circuit a few years back and lived to tell the tale, uncensored.

Doug grew up in Louisiana. His Saturday TV heroes were Hulk Hogan, Chuck Norris, and, of course, Arnold. He thought these guys and their bodies were something to behold, and he wanted to be like them. He excelled at wrestling in high school, and in college at Auburn he found a sense of community at the gym. He saw immediate results from lifting weights, and inspired by *Pumping Iron*, Doug managed to win the Mr. Natural Teenage Louisiana competition. The key word in that title is "natural," meaning no steroids—nothing except hard work, good genes, and optimized nutrition.

The world of competitive bodybuilding has essentially created two leagues: one for the dopers and one for the non-dopers. Doug offered this framework: "There are the *natural* naturals, the naturals, and the unnaturals," the implication being that "natural" is something of a matter of degree—or may quite literally be a matter of dose. *Natural naturals* use protein powder or standard dietary supplements; *naturals* use testosterone, or even trenbolone (veterinary-grade testosterone), although at a lower dose than some others; and *unnaturals* will do whatever it takes—including taking whatever they can—to reach their physical limits. Some in the community call the unnaturals "freaks." Drugs have pushed the sport beyond art and off into the downslope of the bell-shaped curve. Beauty is out; freaks are in. As Doug put it, "Every year they keep getting bigger, but people [spectators] seem to be most interested in seeing them." Aristotle be damned—for this crowd, beauty is defined by size; more specifically, the size of the audience.

Doug somewhat reluctantly decided to try "test"—the community's name for testosterone—and got "geared up" by a friend. At first, the whole thing seemed a little suspicious. "I'm pretty sure the first steroids I used were for veterinary use," he confessed to me. His first clue? "They had pictures of cattle on the bottle."

Yep, I would be suspicious of that, too. But it is a very real drug: Trenbolone, or "tren," is given to beef cattle in the form of a pellet behind the ear when the cow is at the feedlot. It bulks up the cattle, increasing their muscle mass by between 20 and 40 percent. That's a lot of beef, and there's a pretty hefty return on the investment into trenbolone for cattle. It is perfectly legal, and about 80 percent of beef cattle in the United States get dosed with it or something similar before slaughter.* But trenbolone is listed as a Schedule III controlled substance in the United States for *human* consumption, and the World Anti-Doping Agency (WADA), in its 2016 report on doping at the Olympics, cited trenbolone as the agent used by the Russian athletes in Sochi who were caught swapping urine samples.

As for Doug, nothing happened with the stuff from the cow bottle. "It was probably fake," he told me, and he's probably right. One of the downsides of delving into an unregulated drug market is that there's no quality control.

"But then I got something real—finally. It was T-cypionate."

Testosterone for supplementation comes in a lot of forms. The forms differ by the composition of a second molecule to which the testosterone is adhered. In many cases, pure testosterone is applied to the skin as a gel. In Doug's case, the testosterone is adhered to a *cypionate* molecule. These varied formulations allow for differences in things such as duration of action and potency. When it comes to potency (the effect per unit of dose) T-cypionate is about one-seventh the potency of trenbolone and about a sixth that of stanazolol, another form that is often used in bodybuilding. T-cypionate is an intramuscular injection, and, as with all testosterone supplements, the effects of any given dose are highly individualized.

---

*Studies done in the river systems downstream of feedlots in Nebraska and Ohio have detected high levels of 17ß-trenbolone, a metabolite of this steroid, in certain fish species. These fish have small gonads and decreased fertility. Interestingly, one of the greatest causes of "endocrine disruption" is not feedlots, but paper and pulp mills. Chemical by-products used in paper production bind to the androgen receptor and stimulate its effects. The results include female fish with male sex characteristics and male-predominant populations. (Ref: Gray, et al., *International Journal of Andrology* 29 (2006), 96–104.)

Doug took T-cypionate once per week for ten weeks. "Your muscles feel like they're going to pop out of your skin," he told me. "And I blew up like a balloon." He also experienced some of the psychological effects of testosterone: "I felt *unbelievable*—not just in the gym, either. I had more confidence everywhere." We've seen the benefits of testosterone supplementation, and those who take moderate doses for "cosmetic" reasons related to bodybuilding can experience those positive effects. But higher doses come with a higher likelihood of side effects and health problems, not to mention a high-complexity dosing regimen that goes far beyond testosterone.

Doug started to follow the step-by-step cycle of dosing—the strategic going on and coming off of the drug. You don't just stop cold turkey, and in fact you need a veritable hormonal soup of support while withdrawing from testosterone. In perhaps a strange twist, most of the supplements used to come off testosterone are associated with female hormonal issues, and *not* using them as you withdraw can cause symptoms similar to those experienced by women suffering from PMS.

First up is Clomid (clomiphene), a drug that women struggling with infertility sometimes take to stimulate ovulation. Clomid prompts the secretion of follicle-stimulating hormone and luteinizing hormone—in men, luteinizing hormone stimulates testosterone production by the testicles. Next is HCG, the hormone most commonly associated with pregnancy—in fact, if a man taking HCG took a pregnancy test, he'd get a positive. HCG's purpose, in Doug's words, is to "blow your nuts back up." The testicles operate like a kind of thermostat for testosterone, and when the brain senses that there is ample testosterone in the body, it shuts down the stimulus hormone (*luteinizing hormone*) that stimulates testicular growth, causing the testicles to shrink. HCG revs this shrinkage back to life.

"And then," Doug tells me, "you have to take Nolvadex—to prevent your breasts from growing." Nolvadex is tamoxifen, a drug used in the treatment of breast cancer. It blocks the estrogen receptor and is analogous to some of the treatments we use in prostate cancer. As testosterone stimulates prostate cancer, estrogen can stimulate breast cancer, and tamoxifen and another class of drugs called the *aromatase inhibitors* limit the ability of estrogen to activate both cancer *and* normal breast-tissue growth; it does the same in bodybuilders who take high-dose testosterone supplements, which cause them to have high levels of estrogen in their blood, as testosterone can be converted into estrogen by the enzyme aromatase.

As an aside, good, old-fashioned *natural* weight lifting can have hormonal effects of its own, and it might be especially beneficial for cancer patients. Exercise has already been shown to reduce fatigue and improve quality of life in cancer patients, and current studies are focusing on whether resistance exercise like weight lifting, which temporarily raises testosterone levels, has the power to chronically increase levels of that hormone—and others, like cortisol, that fight inflammation. We at UCSF are currently doing a clinical trial looking at the effects of weight lifting in patients with prostate cancer who are on androgen-deprivation therapy.

Doug, now thirty-nine, is still into bodybuilding, but is no longer doping. "I got more mature, lost the motivation, and wanted to slim down," he says. Today he's a suburban father and sales associate, and reports no lasting effects from the time he spent taking high-dose testosterone. He was able to move on, but, as we will see, not everyone is so lucky.

'ROIDS AND RAGE?

"When I was on 'test' I was horny all the time, but I never had any behavioral issues. I would certainly see guys who had a problem, but they were assholes. They would use test and be more of an asshole—but I never saw a nice guy convert into a werewolf."

Doug's observation is supported by research indicating that exogenous testosterone administration in men will enhance aggressiveness only in those with baseline aggressive personality traits. Psychologists showed that men with high baseline dominance and low self-control will act more aggressively after receiving testosterone, and in previous chapters we've seen what problems that can cause. The amygdala, which regulates our emotions, has the highest concentrations of androgen receptors of any brain region, and AR density in the amygdala can actually increase as a result of chronic androgen stimulus, as can the size of the amygdala itself.[120] A larger amygdala has also been associated with conditions such as anxiety disorders.

MRI scans can evaluate in real time not just which parts of the brain are receiving preferential blood flow but also variations in the metabolism of brain cells. Brain-mapping studies of long-term steroid users show that various network connections can be compromised. For example, the connection of the amygdala to the superior frontal gyrus—a region associated with emotional regulation—is drastically reduced in current steroid users, compared with past users and those who have never used.

The behavioral implications are many, as described in scientific papers on bodybuilders that compare both current and former steroid users with controls who never used these drugs. One recent study from Norway reported that while many of the "lost connections" in the brain will return over time, both current and former steroid users report higher levels of anxiety, rule-breaking behavior, and other psychological issues. Not surprisingly, those with a greater lifetime dose show more significant connectivity reductions than those with more moderate lifetime exposure.[121] So yes, steroid use will quite literally change the structure of the brain—and not in a good way.

Consider the above in light of the tragic story of Robert Bales.

According to a 2013 article in the *New Republic*, only about 2 percent of those in the military admit to using anabolic steroids.[122] However, given the usual problems of self-reporting, and the potential for disciplinary action, it is likely that the numbers are significantly higher.

At first glance, I'm tempted to give this use a pass. These are individuals who put their lives on the line, and it's not hard to imagine a scenario in which being faster and stronger—perhaps to escape from danger, perhaps to carry an injured colleague to safety—can mean the difference between life and death. This isn't Lance Armstrong doping in the Tour de France or Ben Johnson doping at the Olympics; this is war, and the threats are real. If it were me, I'd take any performance enhancement I could get.

But then, just because something is done for good reasons doesn't mean there won't be negative side effects, and some wonder whether adding testosterone to the brains of those already under the stress of deployment and combat may be too much. When the body and mind are pushed to, and maybe beyond, their limits of normal resilience, further fueling the fragility of that condition can be deadly. This may be what happened during the Kandahar massacre, one of the darkest moments in our nearly two decades of fighting in Afghanistan.

Early on the morning of March 11, 2012, the on-duty officer at Village Stability Platform Belambay in the Kandahar Province of southern Afghanistan was informed by Afghan National Army counterparts that an American soldier had entered their compound, grunted "Hello, how are you?" and then reversed course and slipped out again. In response, American forces conducted a head count and discovered that Staff Sergeant Robert Bales was unaccounted for. Bales, a thirty-nine-year-old former stockbroker from the Cincinnati suburbs, was in his *eleventh* year of active military service, having enrolled in the wake of the attacks on September 11, 2001. The head count was at 3:20 AM, and at 4:10 AM a solitary figure was spotted approaching the American compound on foot. A patrol sped out to the gate and confronted the man, who was wearing camo pants and a T-shirt. It was Bales. The officer in charge noticed blood splatter on Bales's clothing and took him into custody. Bales calmly requested permission to clean himself up. It was clear to those at the gate that he was uninjured, and the blood didn't appear to be his. Bales was taken into custody on suspicion that he had committed a crime.

A few hours later the rising sun illuminated a gruesome scene. Bales had killed sixteen unarmed civilians and wounded six more. According to army transcripts, Bales, after an evening of drinking (which was not allowed on forward combat bases), slipped out of the compound armed with automatic weapons and, in an inexplicable burst of violence, randomly and brutally killed or wounded almost two dozen men, women, and children in two nearby villages. Bales confessed to the crime.

In the investigation that followed, a search of Bales's bunk revealed a stash of stanazolol, a potent androgenic steroid, tucked under the floorboards. Bales had admitted to and even bragged about taking the steroids, which had been given to him by a Green Beret who had come through the base a few weeks before the massacre.

The popular press looks at this incident and others like it as the tragic consequence of overdeployment, mental fatigue, perhaps emerging mental illness, and even *'roid rage*—in this case the last of these being based on the discovery of the stanazolol. Military officials questioned about the incident later admitted that concern had been growing about the increased use of alcohol and steroids at the base before the massacre.

Experts can't settle on whether "'roid rage" is an actual thing, at least if it's defined as a sudden explosion of violence in a previously nonviolent individual. Studies of elevated testosterone in otherwise normal subjects don't support that this behavior occurs out of the blue in someone who is psychologically healthy, but many scientists and clinicians remain open to the possibility that such outbursts could be fueled by anabolic steroids (and particularly stanazolol) against the backdrop of a psychologically *unhealthy* brain. For Bales, the backdrop was eleven years of deployments, battle fatigue, paranoia, and financial and marital stress back home.

In Bales's case, there were warning signs. His behavior had become aggressive and impulsive in the months leading up to the massacre, and at one point, a month or two before the event, he beat an Afghan contractor who had accidentally hit him with a box he was carrying. Just as there is data that childhood trauma can change the activation of the androgen receptor and result in aggression, it is possible that such intense trauma in an adult could have similar effects.

Bales now sits in the United States Military Detention Facility at Fort Leavenworth, serving a life sentence without the possibility of parole. The unidentified Green Beret who had given him the bottle of stanazolol has been discharged from the army. In his defense statement, the soldier said that he knew it might have been some kind of illegal substance but that he "hadn't really thought about it."[123]

BALANCING BEAUTY AND BEAST

Aristotle and his Mr. Olympia-turned-action-hero/governor protégé would agree on one point: brawn can be beautiful, when in balance. Bigness without it may offend the aesthetes amongst us, but I can live with that. Yet, if one thing stands out amidst all the science and stories of steroids, it is the sobering reality that this balance may be more delicate than we understand, or perhaps than we are willing to admit.

Chapter Twelve

WINSTON NEEDS A CUDDLE: HORMONES AND THE NURTURING ENVIRONMENT THAT SHAPES FAMILIES

D rew's workday lasts up to sixteen hours, and he sometimes works seven days a week. When he's not at work, he's on call—pretty much every night. He didn't get a bonus or a raise last year, and his clients rarely show any appreciation for his effort, often seeming to go out of their way to make his life harder. As a boy growing up on the island of Cyprus, and later as an engineering student in the United States, Drew never dreamed that *this* would be his calling. His wife adores him and calls him her hero, but his traditional Cypriot father struggles with Drew's career choice, trying to talk sense into him whenever Drew and his family visit the home country.

If you haven't guessed it yet, Drew is a stay-at-home dad.

As gender roles in America have shifted in the last several decades, the focus has generally been on the changing role of women, particularly in the workforce. The other side of that coin, however—the increasing numbers of men staying home to keep house and raise children—is equally fascinating, and surprisingly relevant in a discussion of how testosterone shapes both biology and behavior. What does it mean for men, whose very evolution has been tied to physical work outside the home, to take on the domestic, nurturing role once traditionally reserved for women? In 2014, the Pew Research Center estimated that 2 million American men are stay-at-home dads like Drew—almost double the 1.1 million counted in 1989—and today men make up 16 percent of all stay-at-home parents, up from 10 percent in 1989.[124] What accounts for such a sharp rise? Is this solely the result of increased opportunities for women, or are there other factors at play, perhaps even some driven by biology?

One thing is certain: the hyper-virile job market isn't what it once was. We no longer need to kill to survive, manual labor is less common in our society than ever before, and relatively few of us go off to fight in wars, compared with past generations. Virility means different things in mate selection these days, and research has shown that in many ways men with higher testosterone may actually be less desirable mates, as they are more likely to be divorced, get into fights, and have lower socioeconomic status.[45] Statistically, marriages are happier if there is less combined testosterone in the couple.

Still, Drew admits he initially felt a little sheepish whenever someone—especially an older person—asked him what he did for work. "Many thought I was lazy or incompetent," he admitted with a wry chuckle. "I also kept getting [unsolicited] leads on job offers when the kids were really young." Yet, whatever people may have thought about his competence, he notes that none of the comments he received had anything to do with masculinity, virility, or what he was "supposed" to be doing. "Nobody here thinks it is *odd* that I am doing this," he said. Drew figures this may be a function of where he lives—the uber-progressive San Francisco Bay Area. That's probably part of it, but it may also be that stay-at-home dads are becoming less unusual overall.

In Drew's case, it was a simple question of economics. During the two pre-kid years of their marriage, both he and his wife, Jenna, worked well-paying, high-intensity jobs—Drew as a mechanical engineer and Jenna as an executive in the downtown headquarters of an investment bank. As the kids arrived (around the time of the 2008 recession), Drew's career hit a speed bump just as Jenna's accelerated, and together the couple made the pragmatic decision that he would stay home with the kids while she continued her lucrative career. Data compiled in 2015 by the stats-analysis site FiveThirtyEight.com revealed that 38 percent of American wives earn more than their husbands, and in a third of those households, the husband had an income of zero.

A decade later, this arrangement still seems to be working for Drew and Jenna, but not all couples are so lucky. The data compiled by FiveThirtyEight looked not just at the work arrangements of heterosexual married couples, but also at the *quality* of those marriages. The data was conflicting—there was no clear correlation between spousal satisfaction and work status—suggesting, as you might expect, that other variables are in play as well.[125] That said, a 2013 study from the University of Chicago's Booth School of Business revealed that the divorce rate is higher in heterosexual married couples when the wife makes more money than the husband. One particularly interesting detail was that the data was not "continuous," meaning the divorce rate was not any higher when the woman made a lot more money than it was when she made just a little more. In short, the divorce rate increased when the wife was making more money, period, but larger differences didn't make things "worse."[126]

So what does this data tell us about the psychology of breaking traditional gender roles? Some feel strongly that they *shouldn't* be broken, while others argue that couples stand a greater chance of happiness when these stereotypes can be overcome in favor of a more egalitarian society. Indeed, the root of the matter may lie not in the sex of who is making money versus staying home to take care of the children, but in how each person *feels* about the role he or she has taken. These emotional states, as you will see, are only partly a function of social constructs—they are also heavily influenced by hormones. If we could do a more formal study of the biology of this phenomenon, we might find that in *stable* marriages in which the wife makes more money, the male has a lower virility triad (lower testosterone level, less active androgen receptor, lower exposure to fetal testosterone), and is thus less attached to the idea of the man as the breadwinner.

TESTOSTERONE AND FATHERHOOD

The growing acceptance of stay-at-home dads like Drew challenges the notion that the rearing of children should be determined by sex—and yet hormone biology can still tell us quite a bit about why we are the way we are. Let's take a look at what we know about testosterone and fatherhood.

First, the transition to fatherhood is associated with a dramatic reduction in testosterone, down from the high levels used to fuel the dating and mating process. Dating-app developers take note: one study showed that men who went on to have children had higher testosterone levels than those who did not . . . *four years* before the birth of a baby.[127, 128] We've already seen how high testosterone works in attracting a mate, but once the man transitions into nurturing mode, those levels come down.

Considerable attention is paid to the hormonal shifts within mothers and how they affect the growth, development, and nurturing of a newborn, but it's less well known that fathers of newborns also go through a considerable fluctuation during this time, most notably in their levels of testosterone and of oxytocin—the hormone of nurturing and affection.* Testosterone levels are 33 percent lower in fathers of newborns than in non-fathers, while oxytocin levels are more than 25 percent higher.[129] The effect of this new hormonal balance is a greater degree of "paternal investment"—more oxytocin induces fathers to spend more time with their kids and respond more quickly to their needs, whereas more testosterone has the opposite effect. Some studies have shown that higher levels of testosterone, or even simply larger testicles (measured by ultrasound to make it scientific),[130] can make fathers less likely to respond, or simply slower to respond, to the crying of their infant children.

As we saw earlier, administration of intranasal oxytocin will activate the network of the brain that is associated with empathy, and this same mechanism is at work in new fathers.† Fathers who have been given intranasal oxytocin will make closer physical contact with their kids during play and show less frustration with their crying.[131] I recall hearing from the wife of one of my patients that, after he underwent testosterone depletion, he would get down on the floor and play with his grandkids, something he had never done before. Less testosterone, more oxytocin, more bonding.‡ And the bonding behavior, in turn, leads to more oxytocin. Most important, perhaps, is that a parent can

---

*Incidentally, prolactin, testosterone, and oxytocin levels correlate within couples. In other words, as any one of these hormones goes down in the male, it also goes down in the female.[7]

†Intranasal oxytocin has been shown to soften individuals without children who respond harshly (rather than empathetically) to the cry of a stranger's infant. (Think of the last time you were on an airplane with a fussy baby . . .)

‡Nasal oxytocin spray hasn't gone on the market yet, but it might someday. For now, the only way to get it is a natural way, one of which is through physical affection. Cuddle parties have been held for years, and in some cities, you can even hire a professional "cuddler" to get the personal touch you need. Cuddle Con, debuting in 2015 in Portland, Oregon, was the first-ever conference devoted to the art and science of cuddling.

actually "infect" his or her *child* with oxytocin, in a manner of speaking. Oxytocin levels go up in kids who are played with. Scientists have even found that variations in the gene for the oxytocin receptor and another protein called CD38 may accelerate this cross-generational "transfer" of oxytocin.[132] Oxytocin is a feed-forward system, with obvious positive evolutionary influences in humans.*

EMPATHY OVERLOAD

A team of anthropologists at Emory University in Atlanta measured the activity in various brain areas of men hearing a recording of a random baby's cry and correlated it to hormone levels in the blood as well as CAG-repeat length. Their responses were measured by a functional MRI, which shows blood flow in areas of the brain as a marker of how active the area is. Imagine a meteorologist's Doppler map of storm intensity, with more blood flow signifying more activity in that part of the brain. Not surprisingly, the results from the Emory team were that a less-active AR, as defined by longer CAG repeats, is associated with increased activity of the empathy-inducing *anterior insula*. This is the qualitative reverse of the data from the trolley experiment mentioned previously showing that administering testosterone to female subjects makes them less empathetic.[129] This same group did a similar study in which they showed new fathers and non-fathers pictures of either babies or sexually provocative women. In the fathers, the baby pictures elicited significant responses in the brain areas that control emotion processing, whereas those areas weren't activated in the non-fathers. When the sexually provocative pictures were shown, the brain areas

---

*And dogs. A 2015 paper in the journal *Science* reported that when dogs stared into the eyes of their owners, their oxytocin levels went up, and when oxytocin was given to the dogs via nasal spray, they spent more time staring into their owners' eyes.[133]

associated with reward processing lit up in the non-fathers but not in the fathers! Quite literally, it seems, parenting alters our brains. I'll bet many of you already knew that from experience.

But is it possible to be too sensitive?

While virility may prevent some men from engaging in child care as much as they should, those too far on the other end of that spectrum can have problems of their own. Empathetic *over*arousal stresses the brain, and eventually the ability to act is lost in a swirl of overwhelming emotions.[134, 135]

The anterior insula works by degrees, such that low activity is associated with low empathy, moderate activity with moderate empathy, and so on. At the higher margins of this scale—when the anterior insula is extremely active—what emerges is *anxiety*. In fact, this is thought to be one of the mechanisms behind the prevalence of anxiety in mothers during the postpartum period.

In the study from Emory, on fathers, a bell-shaped curve showed this phenomenon at work, and it had a negative correlation to the subjects' feelings of parental investment. In response to the baby crying, parental investment actually went *down* at the higher ranges of anterior insula activity—that is, if a person is too reactive to a situation, *too empathetic*, it can stymie his ability to function.[129] If the crying baby riles you too much, you can't suck it up and be an effective caregiver. It's not that dissimilar from the experiments on moral decision-making we looked at earlier, in which certain individuals were paralyzed when faced with a dilemma—unable to be unequivocal, so to speak. We can hypothesize that these people might register as having low virility triads, and that differences in these variables interact to produce a spectrum of responses to a given situation. As another example, consider an emergency situation, such as someone collapsing in a public place with an apparent heart attack. Some onlookers will spring into rescue mode, but there are usually one or two others who break out into tears or step away from the action, frozen, unable to function. In those who step back, their inability to act might be interpreted as lack of empathy, but in fact it may be the result of too much empathy.*

We hear a lot today about helicopter parenting and the negative effects of coddling our children in a well-meaning effort to protect them from even the possibility of harm, but what is optimal for the kids? They need empathy, for sure, but how much? Do some of us need to do less empathizing and more sucking up when situations call for us to use a little tough love? Just as our bodies need both testosterone and oxytocin in a healthy balance, our kids need us to be both strong and soft in good measure. The other thing they need is our time.

MAKING TIME

The recession of 2008 is usually analyzed in terms of economics, but one day we may look back on it as a telling moment in the history of masculinity and male gender roles. In the last decade, fathers have been found to spend three times as much time in child-rearing and child-care-related activities as they did in the mid-1960s, and although it's still not a ton of time—on average it amounts to 7.3 hours per week—this is a significant change. Two researchers in Maryland, Sandra Hofferth and Yoonjoo Lee, evaluated related data from the American Time Use Survey, sponsored by the US Bureau of Labor Statistics, and found that in the decade from 2003 to 2013 the number of men taking *primary* responsibility for child-rearing also increased substantially.

---

*In a study published in 2011 in the *Journal of Sexual Medicine*, data supported the idea that anxiety is affected by lower testosterone levels and lower AR activity (based on long CAG repeats). In fact, the effects magnify. Those with the longest CAG repeats and the lowest testosterone levels had the highest anxiety levels of the groups studied.[136] Perhaps some individuals suffering from anxiety disorders could benefit from testosterone supplements. Perhaps this anxiety syndrome explains some of the emotional challenges my patients can experience when their testosterone is taken away.

The next question, then, was whether the increase in the quantity of time fathers spent with their children also led to an increase in the quality of time. Hofferth and Lee sought to answer this question, and also wanted to determine the extent to which "gender display" might counterbalance men's increased opportunity to spend time with their children. To explain what they mean by "gender display," imagine if Drew refused to do much child care—even though, as the stay-at-home spouse, he had more opportunity (and responsibility) to do so. This stereotypically male behavior would be considered a strong gender display. Presumably, and theoretically, strong male-gender displays happen when a man feels his masculinity is threatened, in this case by a wife who earns more money and/or the expectation that he now perform "women's work." In a woman, a strong gender display might manifest itself as pretending to enjoy caring for children even if she doesn't. The expectation is that women are "naturally" drawn to being caregivers and mothers, and in cases where that isn't true, the woman might nevertheless feel pressure to play a traditional maternal role.

At one point in our conversation Drew raises an eyebrow as if a revelation has just hit him.

"The problems I face on a day-to-day basis are really no different from those confronted by a mom," he says, "but I don't feel that I am well suited to the task. For the moms I see, much of raising kids seems like an innate skill. For me, and other dads, it's more of a learned skill."

This may be true—or, the fact that we hear fewer women express the same sentiment may be accounted for by gender display, or simple societal pressure. Maybe Drew feels more freedom to express ambiguity about his suitability for a non-classical gender role, just as a woman might feel more comfortable than a man saying she doesn't feel "suited" for some traditionally male activity.

"I'm doing this out of necessity," Drew said in an unsolicited follow-up. "And I'm not one hundred percent sure I'm doing the right thing."

I'm willing to bet he's in good company with millions of other fathers *and* mothers out there.

As the Maryland researchers combed through their survey data, some interesting patterns emerged. Specifically, they didn't find much evidence of an effect from "gender display." Men whose wives worked didn't appear to undergo a reactionary display of defensive masculinity-protecting behaviors. They found that, overall, the amount of time fathers—and not just the unemployed ones—spent with their children increased over the course of the decade during which they were collecting the data.[137] Not only were fathers becoming more *available* to spend time with their children, but they also were presumably more *willing* to do so. So did they overcome their biological and psychological predispositions, or did their hormones actually change their behaviors and emotions?

HAVE YOU HUGGED YOUR PRESIDENT TODAY?

It goes without saying that the need for parental affection and involvement extends well beyond the infancy of a child. One phrase that sticks with me is, "Have You Hugged Your Kids Today?"—a sentiment I saw plastered on many a Midwestern car in my youth. A significant amount of research shows that children raised in supportive, stable environments develop more sophisticated levels of moral control and higher degrees of altruism, and obviously a big part of that is the nurturing and affection they receive from their parents.

A century before those bumper stickers appeared, parental nurturing was at its lowest, and this was, in fact, a defining feature of affluent families of the Victorian era. Such children often saw their parents for only minutes per day—and only with great formality—while spending the bulk of their time under the care of nurses and nannies. Yet the era produced some of the world's greatest men—and, of course, two of its most devastating wars.

Winston Churchill was one such man, and I have long been a fan of his. Having read several of his biographies, I am struck by how his childhood seems to have shaped his later years. He struggled his whole life with the lack of affection he received from both parents (he was reared by nannies and then shipped off to boarding and military schools), and as an adult he once said, "I can count the number of times I was hugged by my mother." Given what we know about his father, Lord Randolph Spencer-Churchill, it's unlikely Winston was *ever* hugged by the patriarch of the family.

Yet, Winston Churchill spent the better part of his adult life defending and striving to equal the greatness he saw in his distant father. In his own writing, which earned him a Nobel Prize in Literature in 1953, Churchill stated, "Famous men are usually the product of an unhappy childhood."

This concept is not lost in modern political history, either. Bill Clinton's father was absent and longed-for during the turbulent Arkansas childhood of the future president, and, likewise, Barack Obama's father was famously distant and uninvolved in his rearing. Obama, in fact, has hinted at where that left him. He said, "There's a wonderful quote that I thought was LBJ's, but I could never verify it: 'Every man is either trying to live up to his father's expectations or make up for his father's mistakes.'"[138] Obama used this quote in his 2006 book *The Audacity of Hope*, and he brought it up again in an interview with the *New York Times* in 2016. This notion obviously continued to loom large in his psyche.

This brings us to the notion of what the anthropologists call *life-history trade-offs*. If you reproduce, that's great, but it comes with a trade-off: you can't be in mating mode (testosterone-driven) if you are in nurturing mode (oxytocin-driven). For the most part, this compromise is evolutionarily advantageous, but at times we may fight against these roles—or our bodies may, by not shifting "modes" in line with circumstances—and that's where trouble can arise.

Sometimes, what pulls a father away from his children is not work outside the home but rather a sort of "sensation seeking"—infidelity, alcoholism, other problematic risky behavior. Testosterone, sensation seeking, and mating behavior exist in a tense relationship with nurturing behavior, and as in all major competitions, both sides can score points, and the lead may change hands several times.

Sensation seeking can be a sign that we have been pulled out of our lane, and if we give in to sensation-seeking urges while we are supposed to be in nurturing mode, disaster may be imminent. Let's be clear: by sensation seeking, we are not talking about taking the kids to Disneyland to ride the roller coasters; we are talking about dangerous, detrimental risk-taking. Promiscuous sexual behavior by a married man is at the top of this list. One of the greatest threats of sensation seeking is that the thrill of risk is often not satisfied but instead drives the individual on to riskier behaviors that push him further and further away from nurturing mode and may drive up testosterone, which of course exacerbates the problem. Because the testosterone system operates in a feed-forward loop, the behaviors can be their own reward and thus self-perpetuate endlessly. Studies of sensation-seeking fathers who continue their sensation-seeking ways after a baby is born show that they maintain higher testosterone levels for longer.[139] Or is it that dads who maintain testosterone levels longer are more likely to engage in sensation-seeking behaviors?

Sensation seeking isn't all bad—new experiences lead to discovery and growth in many cases. For the sake of this discussion, however, when I use the term "sensation seeking," assume I am referring to negative behaviors that would threaten, not augment, family bonding and stability.

The theory of life-history trade-offs[140] proposes that an organism toggles between spending its physiological energies on survival, such as in a time of starvation, winter, or other environmental stresses, or on reproduction. It can't simultaneously devote its energy to both. This theory relates to the *challenge hypothesis*, now a staple of anthropological thinking, originally proposed in 1990 and based on an in-depth study of sparrows by J. C. Wingfield.*[141] The nurturing life of the parent is not only associated with lower testosterone, but also actually *requires* that testosterone go down, according to the theory. On the other hand, mating is highly competitive, requiring male-to-male confrontation and aggressive pursuit of territory. This requires testosterone. Put another way, it *costs* testosterone. Mating can't take place in times when testosterone is required to drive aggression and competition in other areas, like the search for food. Perhaps that's why birds lay eggs in the spring. Schopenhauer may have been onto something when he said there are probably only two impulses in existence: the drive for food and the drive for sex.

Birds are simple, and they have simple needs. But when you consider that this deeply ingrained biological impulse exists in us humans as well, you can perhaps better understand why sometimes the system runs amok and all the testosterone-driven energy that might one day have been spent on hunting for food can get channeled into risky physical and/or sexual behavior. The sensation-seeking father who skips his kid's piano recital to drive around in a fast car with his mistress is displaying breeding behavior, not nurturing behavior.

WHEN IN DOUBT, OUTSOURCE

So how is a nurturing dynamic rescued from sensation-seeking behavior if the normal balance of oxytocin to testosterone is somehow lost

---

*Curious that a guy named Wingfield is a bird researcher. I know a dentist whose last name is Smiley, and several years ago I met an orthopedic surgeon named Bonebrake. Be mindful what your last name is; it may be your destiny.

or compromised? The simple answer is that nature—and especially mothers—will find a way.

The need for testosterone in fatherhood should, by now, be obvious: it drives fathers' provisioning and the protection of both their children and their mates. But throughout our evolution, this same need for provisioning and protection has forced the male to be away from the mothers and the young for extended periods, and sometimes forever, as in the case of death resulting from accidents, fighting, or war. To combat this, mothers have, throughout human development, cultivated networks of co-parents—be they grandparents, older siblings, neighbors, or friends—which together form a community of "allo-mothers," not all of whom are necessarily female. This is the convincing observation and thesis of Sarah Blaffer Hrdy, a primate anthropologist and the author of *Mothers and Others* (2011), a formative book on this topic.[142] Extending this thread beyond what it takes for humans to survive, she posits that this communal child-rearing has played a part in our species' unique ability to understand each other as complex emotional beings. Hrdy's work challenges the notion that the success of nurturing rests on the foundation of a "maternal instinct," and her data suggests that mothering and nurturing are not necessarily automatic. In effect, the data on oxytocin and nurturing mirrors what we've seen with testosterone and winning—oxytocin begets nurturing begets oxytocin, just as testosterone begets winning begets testosterone.

Raising children, argues Hrdy, is best done in a cooperative environment, with the help and input of others. As stay-at-home fathers, no longer away from home provisioning and protecting, Drew and his ilk join this community of allo-mothers, even as the biological mother helps flip the anthropological script by taking on the role of provider. Given our evolution, this arrangement seems more than a bit paradoxical, but in the modern world we can recognize the ways in which both roles contribute to benefit children—and, as long as everyone is happy playing the role undertaken, the ways that both roles benefit the relationships between parents and between parents and their children, not to mention how they benefit society as a whole. As we blur the distinctions between our deeply ingrained gender roles, we make room for everyone to find his or her place. Contrary to what you might think, Drew has plenty of opportunities to feel valued—even for his masculinity—while staying at home with the kids. "The truth is," he added, "I think I've become an important part of the community here because I'm at home. I have a lot of female friends, of course—moms— and they have no problem calling me to help with certain tasks during the day while their husbands are in the city working."

We have learned so much about testosterone and the complicated system it works within, but at its most basic, virility is still largely about competition, winning, and conquest. Competition has been a critical component of our evolution; perhaps competition in the rearing and raising of our young, whether as biological parents, unrelated guardians, mentors, teachers, or coaches, may offer us an opportunity to shape that evolution. Indeed, the acceleration of our evolution came when the need to compete was overcome by (or at least challenged by) the willingness to cooperate. One of the things that makes our species unique is the degree to which we can overcome our biological urges. We are, above all else, complex creatures with the ability to make complex decisions. We are, once and for all, more than just our biology.

CONCLUSION

The paradoxes of virility are in starker relief today than at any other point in history. There is plenty of good news, but also plenty of bad. One state has offered gender-neutral birth certificates (and chances are that others will follow suit), yet a man who defended his penis size in a national televised debate is now the leader of the free world—and no friend to the transgender community. Medicine has made great strides in treating prostate cancer by lowering a man's testosterone level to within the range common for women, yet athletically gifted women are banned from competition because their natural hormone levels fall within the range common for men. Taking supplemental testosterone can make us healthier, and maybe even make us live longer, but it may detract from our ability to enjoy the benefits of either.

Our understanding of this molecular system is just beginning to unfold, and we must remember that it is but one node in the complex matrix that is human biology and human life. My biggest challenge in writing this book has been to avoid the pitfall of reductionism—the fallacy of assuming that, because there are abnormalities in the system, be it an aberration of the androgen receptor, or a low serum-testosterone level, or any other kind of irregularity, such findings are causative, or even the major lever influencing an outcome, be it one related to behavioral or to physical health.

I have presented in this book two triads for your consideration. The first is the virility triad, showing that the true effect of testosterone on humans flows through three distinct channels: testosterone levels in the blood, the androgen receptor and its many variances, and the influence of fetal testosterone. The second triad is the relationship of our bodies, brains, and society, and how they seem to interact and play off one another.

I have seen so many men go through the process of losing their testosterone that observing its effects has become almost intuitive, like a muscle memory. I see virility and its perils in everything, beginning with my morning shave and all throughout the day until I go to sleep. The real challenge has been accurately describing what I notice and what I know in a way that will help other people see and understand it, too. Even after decades in the medical field, assembling all this data—the robust, the weak, and the misleading—has changed me. I think I understand my patients better now, and I also understand myself better—my drives, my moods, and maybe some of the opaque mechanisms behind consciousness. This stuff isn't taught in medical school, or in the training one receives for a career in oncology, urology, or even psychiatry, and I am grateful to have learned it, and to be able to share it with a wider audience.

Although not all of the sources I used have been referenced in these pages, I have read a mountain of research and in doing so have arrived at two compelling realizations: (1) much of the research is insufficient to prove causation, although it often makes compelling arguments for association; and, most important, (2) we are at the beginning—in fact, the very beginning—of our understanding of the testosterone system.

Ever since, and in fact well before, the German chemist Adolf Butenandt isolated what we now call testosterone from 15,000 liters of urine collected from the Berlin Police Department in the 1930s, the

thoughtful and the curious have struggled to make sense of this molecule and the biological system it drives. At the heart of our knowledge
is the fact that while virility gives life amazing color and dynamism,
it can also extract a cost. That's a paradox we'll just have to live with.

NOTES

1. Schroeder, J.P. and M.G. Packard, "Role of Dopamine Receptor Subtypes in the Acquisition of a Testosterone Conditioned Place Preference in Rats," *Neurosci Lett*, 2000. 282(1–2): p. 17–20.
2. Li, C.Y., et al., "Fighting Experience Alters Brain Androgen Receptor Expression Dependent on Testosterone Status," *Proc Biol Sci*, 2014. 281(1796): p. 20141532.
3. Dabbs, J.M., E.C. Alford, and J.A. Fielden, "Trial Lawyers and Testosterone: Blue-Collar Talent in a White-Collar World," *Journal of Applied Social Psychology*, 1998. 28(1): p. 84–94.
4. Beheim, B. (2002). "The Tsimane Health and Life History Project." Retrieved November, from http://www.unm.edu/~tsimane/
5. Coates, J.M. and J. Herbert, "Endogenous Steroids and Financial Risk Taking on a London Trading Floor," *Proc Natl Acad Sci USA*, 2008. 105(16): p. 6167–72.
6. Bernhardt, P.C., et al., "Testosterone Changes During Vicarious Experiences of Winning and Losing Among Fans at Sporting Events," *Physiol Behav*, 1998. 65(1): p. 59–62.
7. Stearns, S.C., "Life History Evolution: Successes, Limitations, and Prospects," *Naturwissenschaften*, 2000. 87(11): p. 476–86.
8. Overskeid, G., "Power and Autistic Traits," *Front Psychol*, 2016. 7: p. 1290.
9. Baron-Cohen, S., "Autism, Hypersystemizing, and Truth," *Q J Exp Psychol* (Hove), 2008. 61(1): p. 64–75.
10. Manning, J.T., et al., "The Ratio of 2nd to 4th Digit Length: A Predictor of Sperm Numbers and Concentrations of Testosterone, Luteinizing Hormone and Oestrogen," *Hum Reprod*, 1998. 13(11): p. 3000–4.

11. Montoya, E.R., et al., "Testosterone Administration Modulates Moral Judgments Depending on Second-to-Fourth Digit Ratio," *Psychoneuroendocrinology*, 2013. 38(8): p. 1362–9.
12. Joel, D., et al., "Sex Beyond the Genitalia: The Human Brain Mosaic," *Proc Natl Acad Sci USA*, 2015. 112(50): p. 15468–73.
13. Baron-Cohen, S. and S. Wheelwright, "The Empathy Quotient: An Investigation of Adults with Asperger Syndrome or High Functioning Autism, and Normal Sex Differences," *J Autism Dev Disord*, 2004. 34(2): p. 163–75.
14. Baron-Cohen, S., "The Extreme Male Brain Theory of Autism," *Trends Cogn Sci*, 2002. 6(6): p. 248–254.
15. Nordahl, C.W., et al., "Brain Enlargement Is Associated with Regression in Preschool-Age Boys with Autism Spectrum Disorders," *Proc Natl Acad Sci USA*, 2011. 108(50): p. 20195–200.
16. Tordjman, S., et al., "Plasma Androgens in Autism," *J Autism Dev Disord*, 1995. 25(3): p. 295–304.
17. Knickmeyer, R., et al., "Fetal Testosterone, Social Relationships, and Restricted Interests in Children," *J Child Psychol Psychiatry*, 2005. 46(2): p. 198–210.
18. Auyeung, B., et al., "Fetal Testosterone and Autistic Traits in 18 to 24-Month-Old Children," *Mol Autism*, 2010. 1(1): p. 11.
19. Auyeung, B., et al., "Fetal Testosterone Predicts Sexually Differentiated Childhood Behavior in Girls and in Boys," *Psychol Sci*, 2009. 20(2): p. 144–8.
20. Auyeung, B., et al., "Fetal Testosterone and Autistic Traits," *Br J Psychol*, 2009. 100(Pt 1): p. 1–22.
21. Baron-Cohen, S., et al., "The 'Reading the Mind in the Eyes' Test Revised Version: A Study with Normal Adults, and Adults with Asperger Syndrome or High-Functioning Autism," *J Child Psychol Psychiatry*, 2001. 42(2): p. 241–51.
22. Van Honk, J., et al., "Testosterone Administration Impairs Cognitive Empathy in Women Depending on Second-to-Fourth Digit Ratio," *Proc Natl Acad Sci USA*, 2011. 108(8): p. 3448–52.
23. Coates, J.M., M. Gurnell, and A. Rustichini, "Second-to-Fourth Digit Ratio Predicts Success Among High-Frequency Financial Traders," *Proc Natl Acad Sci USA*, 2009. 106(2): p. 623–8.
24. Holtfrerich, S.K., et al., "Endogenous Testosterone and Exogenous Oxytocin Modulate Attentional Processing of Infant Faces," *PLoS One*, 2016. 11(11): p. e0166617.

25. Auyeung, B., et al., "Oxytocin Increases Eye Contact During a Real-Time, Naturalistic Social Interaction in Males with and Without Autism," *Transl Psychiatry*, 2015. 5: p. e507.

26. Ruytjens, L., et al., "Functional Sex Differences in Human Primary Auditory Cortex," *Eur J Nucl Med Mol Imaging*, 2007. 34(12): p. 2073–81.

27. Chlebowski, R.T., et al., "Estrogen Plus Progestin and Breast Cancer Incidence and Mortality in Postmenopausal Women," *JAMA*, 2010. 304(15): p. 1684–92.

28. Anderson, G.L., et al., "Effects of Conjugated Equine Estrogen in Postmenopausal Women with Hysterectomy: The Women's Health Initiative Randomized Controlled Trial," *JAMA*, 2004. 291(14): p. 1701–12.

29. Mallen, A., "I Lived Like a Man for a Couple of Weeks. It Helped Me Understand My Husband," *Washingtonpost.com*, June 30, 2015. Retrieved from https://www.washingtonpost.com/posteverything/wp/2015/06/30/what-i-learned-when-i-lived-like-a-man-for-a-couple-of-weeks/

30. Merritt, P., et al., "Administration of Dehydroepiandrosterone (DHEA) Increases Serum Levels of Androgens and Estrogens but Does Not Enhance Short-Term Memory in Post-Menopausal Women," *Brain Res*, 2012. 1483: p. 54–62.

31. Pearcey, S.M., K.J. Docherty, and J.M. Dabbs, Jr., "Testosterone and Sex Role Identification in Lesbian Couples," *Physiol Behav*, 1996. 60(3): p. 1033–5.

32. Meyer-Bahlburg, H.F., et al., "Sexual Orientation in Women with Classical or Non-Classical Congenital Adrenal Hyperplasia as a Function of Degree of Prenatal Androgen Excess," *Arch Sex Behav*, 2008. 37(1): p. 85–99.

33. Agrawal, R., et al., "Prevalence of Polycystic Ovaries and Polycystic Ovary Syndrome in Lesbian Women Compared with Heterosexual Women," *Fertil Steril*, 2004. 82(5): p. 1352–7.

34. Bosinski, H.A., et al., "A Higher Rate of Hyperandrogenic Disorders in Female-to-Male Transsexuals," *Psychoneuroendocrinology*," 1997. 22(5): p. 361–80.

35. Elaut, E., et al., "Relation of Androgen Receptor Sensitivity and Mood to Sexual Desire in Hormonal Contraception Users," *Contraception*, 2012. 85(5): p. 470–9.

36. Gelstein, S., et al., "Human Tears Contain a Chemosignal," *Science*, 2011. 331(6014): p. 226–30.

37. Beauvoir, S.de, C. Borde, and S. Malovany-Chevallier, *The Second Sex*, 2009. London: Jonathan Cape. xxv, p. 822.

38. Stoljar, N., "The Politics of Identity and the Metaphysics of Diversity," in *Proceedings of the 20th World Congress of Philosophy*, 2000. Bowling Green, OH: Bowling Green State University.

39. Auchus, R.J., et al., "Abiraterone Acetate to Lower Androgens in Women with Classic 21-Hydroxylase Deficiency," *J Clin Endocrinol Metab*, 2014. 99(8): p. 2763–70.

40. Longman, J., "Understanding the Controversy Over Caster Semenya," *The New York Times*, 2016.

41. Longman, J., "East German Steroids' Toll: 'They Killed Heidi,'" *New York Times*, 2004.

42. Olson, W., "GDR Athletes Sue Over Steroid Damage," *BBC News*, March 13, 2005. Retrieved from http://news.bbc.co.uk/2/hi/europe/4341045.stm

43. Plato and R. Waterfield, *Republic*, 1993. New York: Oxford University Press, lxxii, p. 475.

44. Van Anders, S.M. and N.V. Watson, "Testosterone Levels in Women and Men Who Are Single, in Long-Distance Relationships, or Same-City Relationships," *Horm Behav*, 2007. 51(2): p. 286–291.

45. Booth, A. and J.M. Dabbs, "Testosterone and Men's Marriages," *Social Forces*, 1993. 72(2): p. 463–477.

46. Eisenegger, C., J. Haushofer, and E. Fehr, "The Role of Testosterone in Social Interaction," *Trends Cogn Sci*, 2011. 15(6): p. 263–71.

47. Dabbs, J.M. and M.G. Dabbs, *Heroes, Rogues, and Lovers: Testosterone and Behavior*, 2000. New York: McGraw-Hill. xvii, p. 284.

48. Edelstein, R.S., et al., "Dyadic Associations Between Testosterone and Relationship Quality in Couples," *Horm Behav*, 2014. 65(4): p. 401–7.

49. Gettler, L.T., et al., "The Role of Testosterone in Coordinating Male Life History Strategies: The Moderating Effects of the Androgen Receptor CAG Repeat Polymorphism," *Horm Behav*, 2017. 87: p. 164–175.

50. Brizendine, L., *The Male Brain*, 2010. New York: Broadway Books. xxi, p. 271.

51. Shores, M.M., et al., "A Randomized, Double-Blind, Placebo-Controlled Study of Testosterone Treatment in Hypogonadal Older Men with Subthreshold Depression (Dysthymia or Minor Depression)," *J Clin Psychiatry*, 2009. 70(7): p. 1009–16.

52. Chao, H.H., et al., "Effects of Androgen Deprivation on Cerebral Morphometry in Prostate Cancer Patients—An Exploratory Study," *PLoS One*, 2013. 8(8): p. e72032.

53. Rosario, E.R., et al., "Brain Levels of Sex Steroid Hormones in Men and Women During Normal Aging and in Alzheimer's Disease," *Neurobiol Aging*, 2011. 32(4): p. 604–13.

54. Nead, K.T., et al., "Androgen Deprivation Therapy and Future Alzheimer's Disease Risk," *J Clin Oncol*, 2016. 34(6): p. 566–71.

55. Beck, M., "Study Warns of Alzheimer's Risk in Some Prostate-Cancer Drugs Men Taking Testosterone-Blocking Drugs Have Nearly Twice the Risk of Developing Disease as Those Using Other Treatments," *Wall Street Journal*, 2015.

56. Pike, C.J., et al., "Protective Actions of Sex Steroid Hormones in Alzheimer's Disease," *Front Neuroendocrinol*, 2009. 30(2): p. 239–58.

57. Raber, J., "AR, ApoE, and Cognitive Function," *Horm Behav*, 2008. 53(5): p. 706–15.

58. Cherrier, M.M., et al., "Testosterone Improves Spatial Memory in Men with Alzheimer Disease and Mild Cognitive Impairment," *Neurology*, 2005. 64(12): p. 2063–8.

59. Orengo, C., et al., "Do Testosterone Levels Relate to Aggression in Elderly Men with Dementia?" *J Neuropsychiatry Clin Neurosci*, 2002. 14(2): p. 161–6.

60. Xing, Y., et al., "Associations Between Sex Hormones and Cognitive and Neuropsychiatric Manifestations in Vascular Dementia (VaD)," *Arch Gerontol Geriatr*, 2013. 56(1): p. 85–90.

61. Luchetti, S., et al., "Neurosteroid Biosynthetic Pathways Changes in Prefrontal Cortex in Alzheimer's Disease," *Neurobiol Aging*, 2011. 32(11): p. 1964–76.

62. Lehmann, D.J., et al., "Association of the Androgen Receptor CAG Repeat Polymorphism with Alzheimer's Disease in Men," *Neurosci Lett*, 2003. 340(2): p. 87–90.

63. Lehmann, D.J., et al., "The Androgen Receptor CAG Repeat and Serum Testosterone in the Risk of Alzheimer's Disease in Men," *J Neurol Neurosurg Psychiatry*, 2004. 75(1): p. 163–4.

64. Anonymous, "Effects of Sexual Activity on Beard Growth in Man," *Nature*, 1970. 226(5248): p. 869–70.

65. Narad, S., et al., "Hormonal Profile in Indian Men with Premature Androgenetic Alopecia," *Int J Trichology*, 2013. 5(2): p. 69–72.

66. Prodi, D.A., et al., "EDA2R Is Associated with Androgenetic Alopecia," *J Invest Dermatol*, 2008. 128(9): p. 2268–70.

67. Botchkarev, V.A. and M.Y. Fessing, "EDAR Signaling in the Control of Hair Follicle Development," *J Investig Dermatol Symp Proc*, 2005. 10(3): p. 247–51.

68. Richards, J.B., et al., "Male-Pattern Baldness Susceptibility Locus at 20p11," *Nat Genet*, 2008. 40(11): p. 1282–4.

69. Hamilton, J.B., "Patterned Loss of Hair in Man; Types and Incidence," *Ann N Y Acad Sci*, 1951. 53(3): p. 708–28.

70. Casto, A.M., et al., "A Tale of Two Haplotypes: The EDA2R/AR Intergenic Region Is the Most Divergent Genomic Segment Between Africans and East Asians in the Human Genome," *Hum Biol*, 2012. 84(6): p. 641–94.
71. I.S.o.H.R.S, "New International Society of Hair Restoration Surgery Survey Finds the Number of Hair Restoration Patients Worldwide," *Reuters*, 2009. Retrieved from http://www.reuters.com/article/2009/06/24
72. Chu, B., "Bill Gates: Why Do We Care More About Baldness Than Malaria?" *The Independent*, 2013.
73. Irwig, M.S., "Depressive Symptoms and Suicidal Thoughts Among Former Users of Finasteride with Persistent Sexual Side Effects," *J Clin Psychiatry*, 2012. 73(9): p. 1220–3.
74. Frye, C.A., "Some Rewarding Effects of Androgens May Be Mediated by Actions of Its 5Alpha-Reduced Metabolite 3Alpha-Androstanediol," *Pharmacol Biochem Behav*, 2007. 86(2): p. 354–67.
75. Giatti, S., et al., "Effects of Subchronic Finasteride Treatment and Withdrawal on Neuroactive Steroid Levels and Their Receptors in the Male Rat Brain," *Neuroendocrinology*, 2016. 103(6): p. 746–57.
76. Di Loreto, C., et al., "Immunohistochemical Evaluation of Androgen Receptor and Nerve Structure Density in Human Prepuce from Patients with Persistent Sexual Side Effects After Finasteride Use for Androgenetic Alopecia," *PLoS One*, 2014. 9(6): p. e100237.
77. Cauci, S., et al., "Androgen Receptor (AR) Gene (CAG)n and (GGN)n Length Polymorphisms and Symptoms in Young Males with Long-Lasting Adverse Effects After Finasteride Use Against Androgenic Alopecia," *Sex Med*, 2017. 5(1): p. e61–e71.
78. Zhou, C.K., et al., "Relationship Between Male Pattern Baldness and the Risk of Aggressive Prostate Cancer: An Analysis of the Prostate, Lung, Colorectal, and Ovarian Cancer Screening Trial," *J Clin Oncol*, 2014. 33(5): p. 419–425.
79. Lesko, S.M., L. Rosenberg, and S. Shapiro, "A Case-Control Study of Baldness in Relation to Myocardial Infarction in Men," *JAMA*, 1993. 269(8): p. 998–1003.
80. Su, L.H., et al., "Association of Androgenetic Alopecia with Mortality from Diabetes Mellitus and Heart Disease," *JAMA Dermatol*, 2013. 149(5): p. 601–6.
81. Muscarella, F. and M. R. Cunningham, "The Evolutionary Significance and Social Perception of Male Pattern Baldness and Facial Hair," *Ethology and Sociobiology*, 1996. 17: p. 99–117.

82. Truman, Jennifer L., and Lynn Langton, "Criminal Victimization, 2014," *Bureau of Justice Statistics*, 2015. NCJ 248973.

83. Baumeister, R.F., *Is There Anything Good About Men? How Cultures Flourish by Exploiting Men*, 2010. New York: Oxford University Press.

84. Ellis, L. and A.W. Hoskin, "Criminality and the 2D:4D Ratio: Testing the Prenatal Androgen Hypothesis," *Int J Offender Ther Comp Criminol*, 2015. 59(3): p. 295–312.

85. Rajender, S., et al., "Reduced CAG Repeats Length in Androgen Receptor Gene Is Associated with Violent Criminal Behavior," *Int J Legal Med*, 2008. 122(5): p. 367–72.

86. Cheng, D., et al., "Association Study of Androgen Receptor CAG Repeat Polymorphism and Male Violent Criminal Activity," *Psychoneuroendocrinology*, 2006. 31(4): p. 548–52.

87. Gottschall, J.A. and T.A. Gottschall, "Are Per-Incident Rape-Pregnancy Rates Higher than Per-Incident Consensual Pregnancy Rates?" *Hum Nat*, 2003. 14(1): p. 1–20.

88. Ellis, L., "Sex, Status, and Criminality: A Theoretical Nexus," *Soc Biol*, 2004. 51(3–4): p. 144–60.

89. Moore, L., "Rep. Todd Akin: The Statement and the Reaction," *New York Times*, August 20, 2012.

90. Klint, T., et al., "Hormonal Correlates of Male Attractiveness During Mate Selection in the Mallard Duck (Anas Platyrhynchos)," *Horm Behav*, 1989. 23(1): p. 83–91.

91. Kochanek, K.D., et al., "Deaths: Final Data for 2014, in National Vital Statistics Reports," *US Department of Health and Human Services Centers for Disease Control and Prevention National Center for Health Statistics National Vital Statistics System*, 2016.

92. Perrin, J.S., et al., "Growth of White Matter in the Adolescent Brain: Role of Testosterone and Androgen Receptor," *J Neurosci*, 2008. 28(38): p. 9519–24.

93. Raznahan, A., et al., "Longitudinally Mapping the Influence of Sex and Androgen Signaling on the Dynamics of Human Cortical Maturation in Adolescence," *Proc Natl Acad Sci USA*, 2010. 107(39): p. 16988–93.

94. Butovskaya, M.L., et al., "Androgen Receptor Gene Polymorphism, Aggression, and Reproduction in Tanzanian Foragers and Pastoralists," *PLoS One*, 2015. 10(8): p. e0136208.

95. Wacker, D.W., et al., "Dehydroepiandrosterone Heightens Aggression and Increases Androgen Receptor and Aromatase mRNA Expression in the Brain of a Male Songbird," *J Neuroendocrinol*, 2016. 28(12).

96. Newman, A.E. and K.K. Soma, "Aggressive Interactions Differentially Modulate Local and Systemic Levels of Corticosterone and DHEA in a Wild Songbird," *Horm Behav*, 2011. 60(4): p. 389–96.

97. Lucas, G., "Wilson to Sign Castration Bill Today," *San Francisco Chronicle*, 1996.

98. Bund, J.M., "Did You Say Chemical Castration?" *University of Pittsburgh Law Review*, 1997. 59(1): 157–192.

99. Schober, J.M., et al., "Leuprolide Acetate Suppresses Pedophilic Urges and Arousability," *Arch Sex Behav*, 2005. 34(6): p. 691–705.

100. Schmucker, M. and F. Losel, "Does Sexual Offender Treatment Work? A Systematic Review of Outcome Evaluations," *Psicothema*, 2008. 20(1): p. 9–10.

101. Truman, Jennifer L., and Lynn Langton, "Criminal Victimization, 2014," *Bureau of Justice Statistics*, 2015. NCJ 248973.

102. Macrae, A. D., *W.B. Yeats: A Literary Life*, 1995. Basingstoke: Macmillan.

103. Rubin, M., "Yeats's 'Mystical Marriage,'" *Christian Science Monitor*, January 19, 1993. Retrieved from www.csmonitor.com/1993/0119/19131.html

104. Dunbar, N., C.-E. Brown-Séquard, and G. Variot, *The 'Elixir of Life'*, 1889, Boston: J. G. Cupples Company.

105. Brown-Sequard, C.E., "Note on the Effects Produced on Man by Subcutaneous Injections of a Liquid Obtained from the Testicles of Animals," *Lancet*, 1889. 134(3438): p. 105–107.

106. Mayberg, H.S., et al., "The Functional Neuroanatomy of the Placebo Effect," *Am J Psychiatry*, 2002. 159(5): p. 728–37.

107. Kaptchuk, T.J., "Powerful Placebo: The Dark Side of the Randomised Controlled Trial," *Lancet*, 1998. 351(9117): p. 1722–5.

108. Arnold, M.H., D.G. Finniss, and I. Kerridge, "Medicine's Inconvenient Truth: The Placebo and Nocebo Effect," *Intern Med J*, 2014. 44(4): p. 398–405.

109. Hall, K.T., et al., "Catechol-O-Methyltransferase Val158met Polymorphism Predicts Placebo Effect in Irritable Bowel Syndrome," *PLoS One*, 2012. 7(10): p. e48135.

110. Diatchenko, L., et al., "Genetic Basis for Individual Variations in Pain Perception and the Development of a Chronic Pain Condition," *Hum Mol Genet*, 2005. 14(1): p. 135–43.

111. Snyder, P.J., et al., "Effects of Testosterone Treatment in Older Men," *N Engl J Med*, 2016. 374(7): p. 611–24.

112. Hildebrandt, T., et al., "Exercise Reinforcement, Stress, and Beta-Endorphins: An Initial Examination of Exercise in Anabolic-Androgenic Steroid Dependence," *Drug Alcohol Depend*, 2014. 139: p. 86–92.

113. Abd Alamir, M., et al., "The Cardiovascular Trial of the Testosterone Trials: Rationale, Design, and Baseline Data of a Clinical Trial Using Computed Tomographic Imaging to Assess the Progression of Coronary Atherosclerosis," *Coron Artery Dis*, 2016. 27(2): p. 95–103.

114. Von Drehle, D., "Feeling Deflated? The Low-T Industry Wants to Pump You Up," *Time*, 2014.

115. McGlothlin, J.W., et al., "Natural Selection on Testosterone Production in a Wild Songbird Population," *Am Nat*, 2010. 175(6): p. 687–701.

116. Zak, P.J., et al., "Testosterone Administration Decreases Generosity in the Ultimatum Game," *PLoS One*, 2009. 4(12): p. e8330.

117. Emanuel, Ezekiel, "Why I Hope to Die at 75: An Argument That Society and Families—And You—Will Be Better Off If Nature Takes Its Course Swiftly and Promptly," *The Atlantic*, October 2014.

118. Sutin, A.R. and A. Terracciano, "Perceived Weight Discrimination and Obesity," *PLoS One*, 2013. 8(7): p. e70048.

119. "Perceptions of Perfection Across Borders," *Superdrug.com*, 2015. Retrieved from https://onlinedoctor.superdrug.com/perceptions-of-perfection/

120. Menard, C.S. and R.E. Harlan, "Up-Regulation of Androgen Receptor Immunoreactivity in the Rat Brain by Androgenic-Anabolic Steroids," *Brain Res*, 1993. 622(1–2): p. 226–36.

121. Westlye, L.T., et al., "Brain Connectivity Aberrations in Anabolic-Androgenic Steroid Users," *Neuroimage Clin*, 2017. 13: p. 62–69.

122. Drummond, K., "This Is Your Military on Drugs," *New Republic*, 2013.

123. Ashton, A., "Army Kicked Out Green Beret Who Gave Steroids to Robert Bales Weeks Before Massacre," *News Tribune*, 2014.

124. Livingston, G., "Growing Number of Dads Home with the Kids Biggest Increase Among Those Caring for Family," *Pew Research Center*, 2014.

125. Chalabi, M., "How Many Women Earn More Than Their Husbands?" *FiveThirtyEight*, 2015. Retrieved from www.FiveThirtyEight.com

126. Bertrand, M., E.K. Emir., and J. Pan, "Gender Identity and Relative Income Within Households," *Quarterly Journal of Economics*, 2015. p. 571–614. doi:10.1093/qje/qjv001.

127. Edelstein, R.S., et al., "Prospective and Dyadic Associations Between Expectant Parents' Prenatal Hormone Changes and Postpartum Parenting Outcomes," *Dev Psychobiol*, 2017. 59(1): p. 77–90.

128. Edelstein, R.S., et al., "Prenatal Hormones in First-Time Expectant Parents: Longitudinal Changes and Within-Couple Correlations," *Am J Hum Biol*, 2015. 27(3): p. 317–25.

129. Mascaro, J.S., et al., "Behavioral and Genetic Correlates of the Neural Response to Infant Crying Among Human Fathers," *Soc Cogn Affect Neurosci*, 2014. 9(11): p. 1704–12.

130. Mascaro, J.S., P.D. Hackett, and J.K. Rilling, "Testicular Volume Is Inversely Correlated with Nurturing-Related Brain Activity in Human Fathers," *Proc Natl Acad Sci USA*, 2013. 110(39): p. 15746–51.

131. Weisman, O., R. Feldman, and A. Goldstein, "Parental and Romantic Attachment Shape Brain Processing of Infant Cues," *Biol Psychol*, 2012. 89(3): p. 533–8.

132. Feldman, R., et al., "Parental Oxytocin and Early Caregiving Jointly Shape Children's Oxytocin Response and Social Reciprocity," *Neuropsychopharmacology*, 2013. 38(7): p. 1154–62.

133. Nagasawa, M., et al., "Social Evolution: Oxytocin-Gaze Positive Loop and the Coevolution of Human-Dog Bonds," *Science*, 2015. 348(6232): p. 333–6.

134. Eisenberg, N., "Emotion, Regulation, and Moral Development," *Annual Review of Psychology*, 2000. 51(1): p. 665–97.

135. Batson, C.D. and Powell, A.A., "Altruism and Prosocial Behavior," in *Handbook of Psychology*, 2003. New Jersey: John Wiley & Sons. p. 463–84.

136. Schneider, G., et al., "Sex Hormone Levels, Genetic Androgen Receptor Polymorphism, and Anxiety in 50-Year-Old Males," *J Sex Med*, 2011. 8(12): p. 3452–64.

137. Hofferth, S. and Y. Lee, "Family Structure and Trends in US Fathers' Time with Children, 2003–2013," *Fam Sci*, 2015. 6(1): p. 318–329.

138. Galanes, P., "Table for Three," *New York Times*, 2016.

139. Perini, T., et al., "Sensation Seeking in Fathers: The Impact on Testosterone and Paternal Investment," *Horm Behav*, 2012. 61(2): p. 191–5.

140. Promislow, D.E.L. and P.H. Harvey, "Living Fast and Dying Young: A Comparative Analysis of Life-History Variation Among Mammals," *Journal of Zoology*, 1990. 220: p. 417–437.

141. Wingfield, J.C., "The Challenge Hypothesis: Where It Began and Relevance to Humans," *Horm Behav*, 2016. 92: p. 9–12.

142. Hrdy, S.B., *Mothers and Others: The Evolutionary Origins of Mutual Understanding*, 2009. Cambridge, MA: Belknap Press of Harvard University Press. p. 422.

ACKNOWLEDGMENTS

A huge number of selfless individuals took the time to read my early drafts and offer advice and direction, pull me back from the ledge, and help me avoid complete embarrassment in this process. I only hope that I'll be able to pay those favors forward someday. In particular, huge debts of gratitude are owed to Kelly Parsons, Thomas Goetz, Adam Siegel, Hala Borno, Alicia Morgans, Renee Vollen and friends, the late Tom Perkins, Stan Rosenfeld, Silke Gillessen, and Judy Grimes for their thoughtful input and support on various sections; and to my agent Al Zuckerman for his mentorship, stewardship, and making me feel like a real, live author. To Eric Small and many other colleagues at UCSF who endured my thinking out loud (and may have to continue to do so!). Special shout-out to the thought-invoking tranquility and warmth of the Mill Valley Public Library and to Whitaker Evans of the UCSF library system. Finally, to Jessica, Patrick, Cate, and Elise for being my true north.

INDEX

evolutionary neuroandrogenic (ENA) theory, 120

exogenous testosterone, 43. *see also* testosterone supplementation

"extreme male brain" (EMB) theory, 25

eyes
 emotions conveyed through, 31–32
 eye contact and fetal testosterone, 29

F

face gaze, 28

facial communication, 28–33

facial hair growth, 96–99

FACIT (Functional Assessment of Chronic Illness Therapy), 154n*

fathers
 as caregivers, 192–194
 and hormones, 188–190
 stay-at-home, 185–187

fetal development, 39–40

fetal testosterone, 23–38
 and auditory recognition patterns, 41–42
 and autism, 23–27
 and facial communication, 28–33
 and finger length, 15–16
 and mind-body dualism, 33–38
 and sexual aggression, 118
 and sexual orientation, 48
 women's exposure to, 47–50

finasteride, 103–111

fish, affected by runoff, 156n*, 178n*

Fishbein, Morris, 144

Fisher, Herbert and Zelmyra, 78

5AR (5-alpha reductase), 105, 108–109

FiveThirtyEight.com, 187

flibanserin, 46–47

flutamide, 91

Functional Assessment of Chronic Illness Therapy (FACIT), 154n*

G

GABAergic activity, 109–110

GAMA (gamma-Aminobutyric acid), 109–110

Gates, Bill, 104

gender
 and Alzheimer's disease, 85–86
 and brain size, 25
 differences in, 18, 42–43
 sex vs., 55–57

"gender display," 193

gender roles, 185–188

generosity, 161–162

genetic propensity, for alopecia, 99–103

Gonne, Maud, 137

H

Hadza (tribe), 126

Haire, Norman, 138

hair replacement treatments, 103–104

Hawthorne effect, 146

HCG, 179

health, alopecia as indicator of, 112–114

heart attack, 112–113

Henry VIII, king of England, 112n*

HGH (human growth hormone), 176

high testosterone levels, 107–108

Hofferth, Sandra, 192–193

Hoge, Bill, 128

hormonal therapy
 beneficial side effects of, 20
 permanent effects of, 83
 in treatment of autism, 36–37
 in treatment of prostate cancer, 4–5

hormone replacement therapy (HRT), 43–44

hormones, xiv–xv
 and fatherhood, 188–190
 similar composition of different, xvi

Hrdy, Sarah Blaffer, 198

HRT (hormone replacement therapy), 43–44

human growth hormone (HGH), 176

Hyde-Lees, Georgie, 137n*

I

IAAF (International Association of Athletics Federations), 58–59

IGF (insulin growth factor), 97

impulsivity, 124–125

India, 116